UKRAINE UNDER PERESTROIKA
Ecology, Economics and the Workers' Revolt

Also by David R. Marples

CHERNOBYL AND NUCLEAR POWER IN THE USSR
THE SOCIAL IMPACT OF THE CHERNOBYL DISASTER

UKRAINE UNDER PERESTROIKA
Ecology, Economics and the Workers' Revolt

DAVID R. MARPLES

Research Associate, Canadian Institute of
Ukrainian Studies, University of Alberta, Edmonton

St. Martin's Press   New York

First published in the United States of America in 1991

Printed in Great Britain

ISBN 0–312–06196–X (hardcover)
ISBN 0–312–06197–8 (paperback)

Library of Congress Cataloguing-in-Publication Data
Ukraine under perestroika : ecology, economics, and the workers'
revolt / David R. Marples.
    p.   cm
Includes bibliographical references (p.   ) and index.
ISBN 0–312–06196–X (hardcover). — ISBN 0–312–06197–8 (pbk.)
1. Ukraine—Economic conditions—1945– 2. Perestroika—Ukraine.
3. Pollution—Economic aspects—Ukraine. 4. Chernobyl Nuclear
Accident, Chernobyl, Ukraine, 1986—Social aspects. 5. Ukraine–
–Social conditions—1945– I. Title.
HC337.U5M28   1991
330.947'710854—dc20   91-8110
                                                          CIP

In memory of my daughter,
Nicole Louise Marples
(1986–88)

CONTENTS

LIST OF ILLUSTRATIONS

| ACKNOWLEDGEMENTS

THIS VOLUME dates back to the mid-1980s, when I started work as a Research Analyst at Radio Liberty in Munich. At that time and subsequently, I received great encouragement from Dr. John Eriksen, in his capacity as the Director of the Department of Program Support. More recently, I have been grateful for the constant provision of documents from the considerable resource base of Radio Liberty and express thanks to the current director, S. Enders Wimbush and to the deputy director, Iain Elliot, who first provided me with an outlet for my articles on Ukraine in the bimonthly *Soviet Analyst*.

This book has also benefited from the knowledge of Bohdan Krawchenko, director of the Canadian Institute of Ukrainian Studies, a colleague who not only shares many of my own interests, but one who has cultivated a plethora of useful contacts in Ukraine without which this current work could not have been completed. In addition, I have made extensive use of interviews and talks conducted during various visits to the Soviet Union in 1987–89, the most beneficial of which was a 1989 trip to Kiev and to the Chernobyl nuclear power plant, organized by the Ukrainian Ministry of Foreign Affairs. I express my thanks to the many Soviet citizens who assisted me, in particular to Yurii Shcherbak, Vitalii Karpenko, Yurii Bohayevsky, Yurii Risovanny, Pavel Pokutnyi, Leonid Leshchenko, Dmytro Pavlychko, Yurii Pokalchuk, Oles Pyatak, I.P. Los, Valerii Ingulsky, Andrii Fialko, Solomea Pavlychko; and to the editorial board of the newspaper *Literaturna Ukraina*, and the station *Radio Kiev*.

Many contacts have sent me very useful materials, especially Taras Kuzio of the Ukrainian Press Agency in London, and Roma Hadzewycz, editor of *The Ukrainian Weekly*. I am especially thankful to my main assistant during the final stages of this project, for his proofreading and original ideas: David Duke, a graduate student in the Department of History, University of Alberta. Generous support for Mr. Duke's work was provided by the Ukrainian Canadian Foundation of Taras Shevchenko, Winnipeg. Over the past two years, I have had several useful conversations with Mario Dederichs of *Stern* magazine (Hamburg), and he gave me free access to his excellent collection of photographs, for which I am very grateful. Among the many others who assisted in some way must be singled out Gerard Magennis, Chrystia Freeland, James Dingley, William Werbeniuk, Oleh Ilnytzkyj and J. Holowacz in Canada, and Jane Corbin of BBC's *Panorama* program.

A special thanks goes to my family: to my wife, Lan, and my young sons Carlton and Keelan for tolerating the erratic working hours of their husband and father.

| NOTE ON TRANSLITERATION

For the most part, this book transliterates Ukrainian place names and personalities according to their Ukrainian spelling. An exception has been made for those names that are familiar in English only in their Russian form, such as Chernobyl, Dnieper River, and Kiev. In block quotes from Ukrainian sources, however, the original Ukrainian form is retained: hence Chernobyl becomes Chornobyl and Kiev becomes Kyiv.

For convenience, the use of the cyrillic soft sign (') has been eliminated from names in the text. It is retained in the notes, however.

| INTRODUCTION

WHY STUDY UKRAINE?, is the question that I have been asked most often over the past few years. What would prompt a non-Ukrainian of English background to devote a good part of his academic life to studying that nation? There is no simple answer. My interest in things Ukrainian was sparked at the University of Sheffield in the late 1970s, when my doctoral thesis supervisor—also a non-Ukrainian—upon hearing that I wanted to specialize on some aspect of Soviet economic history, suggested that I choose one of the national republics rather than Russia proper. In his view, there were rather too many graduates in Russian studies and the nationalities field was being neglected. At that time, there was no indicator of what was to happen to the Soviet Union under perestroika. These were the Brezhnev years which today, not entirely accurately, have been termed the "period of stagnation."

I chose Ukraine because I found it the most intriguing of the Soviet republics. It has been a strangely neglected country, rich in resources and agricultural land, and in historical and religious traditions. In the modern period, Ukraine has had only brief glimpses of independence. Independence was akin to a forbidden dream. When Soviet propagandists discussed manifestations of Ukrainian independence, they would equate them with "bourgeois nationalism" or collaboration with the German occupation regime of the World War II years, ignoring the more important declaration of independence and unity of Ukraine in 1918–19. In the West, Ukrainians had long felt frustrated with the lack of attention paid to human rights violations in their homeland. Ukrainian dissidents in the 1960s and 1970s rank alongside Jewish dissidents as the most prominent but nonetheless generally neglected groups. It is often said that Ukraine has a tragic history; it is, one might say, a truism, but I doubt that before the present era many in the West outside the Ukrainian community gave much thought to it.

Even embarking on Ukrainian studies appeared to be an uncertain venture. In the late 1970s and early 1980s, to study Ukraine was to study events in Moscow. At the official, higher levels, very little happened independently. In 1972, Petro Shelest was removed as the Ukrainian Party leader and replaced by the 54-year-old Volodymyr Shcherbytsky, a dull but ruthless figure, who declined to speak his native language and began to russify his republic, firmly repressing any manifestations of Ukrainian culture. Public opinion beyond the sphere of private thought had died. Here was totalitarianism at its most complete, even more so than in the heartland of the Soviet empire, Moscow itself. Moscow spoke and Kiev followed suit like an obedient child which, in turn, was anxious to outdo even its parents in its rigorousness and severity. It is only because Ukrainian public opinion has been revived that one has learned just how completely dead it was in the Shcherbytsky era of 1972–85. Even when change occurred in Moscow, the ageing leadership in Kiev clung to the old ways like a drowning man clings to the side of a lifeboat.

Academically in the West, Ukraine was a subject of interest predominantly to scholars of Ukrainian descent. Ukraine can be compared in size and population with France, and it is of no surprise to anyone that that fair country has been the subject of study by scholars of all nationalities. But non-Ukrainians studying Ukraine could be counted on the fingers of one hand. Moreover, if one was studying Ukraine in the modern period, it was often assumed by outsiders that there must be some political motive involved. Ukrainians have in the past been erroneously and unforgivably characterized as hardline anti-Soviet right-wingers. Part of the fault for such an assumption lies with the KGB, which continued for years to propagate such assertions in the West in the fear that national consciousness among Ukrainians in Ukraine might resurface because of influence from abroad. For my own part, I was in the middle of studies that seemed at times disturbingly peripheral. From 1984–85, I worked in Munich for Radio Liberty, where Roman Solchanyk and I were research analysts on current Ukrainian affairs.

The Gorbachev period, however, saw changes that were so dramatic that the need for this book and its companion volume (by Roman Solchanyk) seems almost self-evident. It is often overlooked in retrospect that there have been several very different stages to the new era. One thinks of glasnost and perestroika as a continual process that began in March 1985 and continued through 1990, with Gorbachev first controlling and then being controlled by events, hurled along a veritable roller coaster with barely one hand on the wheel of his car. He was named, after all, the "Man of the Decade" by *Time* magazine. But between March 1985 and April 1986, how much really changed beyond rhetoric and an unsuccessful campaign against alcoholism? The potential for change was there, but the reality was otherwise. It is scarcely credible to me now that sitting at a discussion in the Novosti Press Agency in late 1987, a leading official there assured me and others that the AIDS virus was started deliberately by the CIA in laboratories in the United States. At that same time, the KGB were rounding up members of the "Glasnost" group (today almost forgotten) and I was almost included in their number.

The change, even in Moscow, came more slowly. The Chernobyl disaster has also been perceived by some as the real beginning of glasnost, which is almost astonishing given the coverup that followed, though it does say much about the power of the Western media to convey a certain image of an event. Only after four years is it widely acknowledged that most of the information about Chernobyl—and virtually all that on its health effects—was classified immediately afterward by Leonid Ilyin and the Soviet health authorities. Gorbachev survived Chernobyl, but the real impetus for change came from the outlying regions of the Union: from the Baltic republics, Armenia, Azerbaidzhan, Moldavia and Georgia, for a variety of reasons. Finally, like a stumbling giant, came Russia itself led by the ebullient and unpredictable Boris Yeltsin. By the end of 1989, it was plain that whatever crises the Soviet Union had endured, there were two that would persist: the nationalities question and the struggling economy. Together they have ensured the lasting success of glasnost and almost certain failure of perestroika.

Even here, Ukraine has fallen behind, or perhaps rather the process began later and more slowly than in other republics. There were few fundamental differences between the demands expressed, say, by the Lithuanian reformers in 1987–88 and those in Estonia and Latvia. There were far more between those of workers in Eastern Ukraine, and those in Odessa or Lviv. Was there one Ukraine, or two, or three? And what of the Ukrainian Popular Movement, or Rukh? Why had it not taken hold of the population as in the Baltic republics? The popular notion was that no matter what happened in the Baltics, the central authorities would take steps to ensure that Ukraine never left the Union. Economically it was the heartland of the country. If Ukraine seceded, then the Soviet Union could not survive, so the argument ran. Thus the nation was being controlled with an iron fist that would tolerate no movement that threatened secession. There was some truth to this notion, but it was also in some ways misleading, for several reasons.

In the first place, the Ukrainian SSR within its presentday territories was a superficial creation that had resulted from the pact between Hitler and Stalin. The incorporation of Western Ukraine in 1939–41 and again in 1944 injected a new degree of national consciousness into Ukrainians. It may have fulfilled an "ageold" desire to reunite the Ukrainian territories into a single entity. But it did not recreate Ukraine as a nation at that time. For one thing, there were linguistic differences: Western Ukrainians spoke Ukrainian; elsewhere, Russian was the dominant language. Stalin attempted to russify Western Ukraine by moving huge numbers of "reliable" cadres from Eastern Ukraine and the Russian Republic into Western Ukraine, but he did not succeed in this task. Covertly perhaps, forced into an underground existence, Ukrainian national consciousness persisted, as did the Ukrainian Catholic Church, officially dissolved in 1944. The Ukrainians had their ancestry in common, but differences remained. The Communist Party was the main unifying factor, or, in short, the two parts of Ukraine were equally repressed under a pitiless totalitarian regime.

Second, and this is a factor that is explored in depth in this book, the country's economic decline was particularly marked in Ukraine. Years of intensive exploitation had taken their toll on the Ukrainian economy, which was run from Moscow and geared to all-Union needs. One can provide an analogy with the decline of northern England, with its coal mines and steelworks. Just as the so-called "boom" of the early Thatcher years in England was dependent upon the southern counties and the city of London, so it would seem that any future economic revival in what is today the Soviet Union must be based on the natural resources of eastern Russia and Siberia. Moscow based the country's development on the European part of the country; now it must turn eastward unless these vast riches become the property of a new and reformed Russian state. I would further posit that contrary to popular belief Ukraine cannot survive alone as an independent state; whether or not this would have been possible in the recent past is debatable. The tremendous debate taking place on economic sovereignty would be enhanced by more input from those who have advocated a very close trading relationship with Russia proper, because herein lies the real key to independent survival.

Thus an assumption that the separatist movement in Ukraine was attacked so viciously because the Union needs or needed Ukraine is, I believe, a false one, though the republic is economically still important. In my view, the most important factor to consider is the ruling hierarchy: the Communist Party of Ukraine, the KGB and those powerful Moscow-based ministries that even in the summer of 1990 still controlled 95 percent of Ukrainian industrial output. When the Western Ukrainian governments adopted the reformist pattern after the spring of 1990, they were the first to throw off this old leadership. What Vyacheslav Chornovil and his peers have accomplished is a revolution at least the equal to what has occurred in Lithuania or Moldavia if only because of the all-pervasive nature of the previous regime. From the party headquarters to the university or institute rectors, Ukrainians were faced with a stultifying, shocking bureaucracy. It was like being in a room without oxygen. Even the Rukh, the Green World, and the Shevchenko Ukrainian Language Society faced a barrage of opposition before they became formally established. One might say that Ukraine had one of the strongest anti-democratic movements in the Soviet Union and that it was directed from above, against the popular will.

Roman Solchanyk and I conceived of these particular volumes in 1987 though they have both been delayed by a series of dramatic events. Volume One has been dependent upon a close, daily monitoring of the Soviet and Ukrainian media, newspapers and scholarly journals over a number of years. It has also benefited from numerous exchanges of opinion with prominent Ukrainians, both inside and outside the Soviet Union.

Ukraine has also become newsworthy, a subject of much discussion in the Western media. Western statespersons such as the Canadian Prime Minister Brian Mulroney, and Margaret Thatcher of Britain, visited Kiev in 1990. The diaspora—those of Ukrainian background living outside Ukraine, in Canada, the United States, Europe, Brazil, Argentina and Australia—has had a major impact on the situation in Ukraine, whether it be through aid for Chernobyl children, international academic exchanges, or material assistance that has enabled the survival and progress of groups such as the Rukh or the Ukrainian Republican Party. Indeed, many Ukrainians in the West have become deeply involved in business and other ventures in Ukraine. In short, from my own perspective, my studies have essentially moved from the outskirts to the mainstream of popular interests, a process that could never have been predicted.

In this volume, my intention has been to raise several questions about the relationship between economic and ecological problems and the development of popular opinion and movements. The focus is on social and economic events that I consider to have been the most significant during the years 1985–90. The goal is to ascertain: how deep is Ukraine's economic recession today? What new information has come to light about the aftermath of Chernobyl and what have been the material, medical and psychological effects of this situation on Ukrainians? How serious is Ukraine's environmental problem? What has been the relationship between the onset of ecological crises and the development of popular actions and protests? All the above topics are discussed in the first five chapters and it will be seen that the rise of patriotism in Ukraine and of the notion that Ukrainians must run their economy has owed much to the conception that the previous administrators had run roughshod over the resources and natural beauty of Ukraine. Perestroika, to many Ukrainians in today's environment, signifies control over the planning and operation of the republican economy.

The final chapter focuses on the coal industry and the coal strike of July 1989, which has had longterm political repercussions in Ukraine. The inclusion of such a chapter serves a dual purpose. First, it provides a vivid illustration of the reaction of the Ukrainian worker to an intolerable economic and social situation. Second, it demonstrates the sentiment and popular opinion in an area of Ukraine not known in the past for political militancy, namely the industrialized eastern oblasts of the republic: Dnipropetrovsk, Donetsk and Luhanske. There are two different popular approaches that are observed in this monograph: the first is that of members of the Green World ecological association (a response largely to Chernobyl, but also to general environmental difficulties in Ukraine), who though incensed with several giant ministries and associations, and even with the pre-September 1989 party leadership, have worked within the existing structure to change the situation. The second approach is that of the strike committees of the Donbass coal miners, which have manifested contempt for the authorities both in Kiev and Moscow, and simply do not believe that change will be possible without a complete change of government.

An underlying theme is evident in this book, which is that of a rift in Ukrainian society, between on the one hand the government ministries, the planners, and the party leaders, and on the other, the general public. In 1987, at least, Ukrainian activists, Rukh members and others, regarded Mikhail Gorbachev as a progressive figure who was striving to rid the Ukrainian leadership of its inertia. There was a link between the growth of popular activism in Ukraine and that in Moscow, with the difference that in Ukraine's case, the impetus came from below. Chernobyl is again an obvious example of this societal division. Ukrainian intellectuals and writers were the first to sense that the story of the tragedy basically had remained untold. Some of these people were party members, and some were among the founders of Rukh. But Rukh would not have developed had the party truly represented the interests of Ukrainians. If the republic was under assault from outside ministries that paid no attention to local conditions, then, many felt, the Ukrainian party leadership should stand up for Ukrainian interests. If there ever was such a patriotic tradition in Ukraine after the 1920s, it did not survive.

The unfortunate fact for Ukrainians is that many of the predicaments explored in this book were clearly avoidable. There was no logical reason why planners would build a canal that linked the Dnipro to the Danube, for example, or why a nuclear power plant would be constructed in a zone known for its earthquakes. These structures were less examples of irrational planning than outright incompetence. But before the development of glasnost, there was no public discussion of economic development, least of all at the republican level. A second and perhaps even more basic point is that much of the damage done from such plans may be irrevocable. How does one replace more than ten thousand dead rivers, for example, or continue existence on soil contaminated with radioactive cesium? When the Green World speaks of the right to "life on earth," it is expressing the very essence of the problem. The grandiose schemes of the past, whether they occurred in the time of Stalin, Khrushchev, Brezhnev or even Andropov have threatened the existence of Ukraine. As we note below, there have already been very clear signs that a new generation of children has been adversely affected by environmental pollution. How is one to stop such pollution?

There is no point in seeking answers to such questions at this juncture. Indeed, one could devote an entire book to the subject of whether it will be possible, in the future, to live in a Ukraine which is not only a sovereign state (this became a reality while the work on this volume was in progress), but one with a clean atmosphere, pure waters, and an economy that is geared to the welfare and health of the citizens rather than to economic growth. Fortunately, the tension between the Super Powers had eased considerably in 1989–90 so that it was at least possible to devote attention to the domestic economy without major diversions. Yet the growing ecological awareness also coincided with the increased divisions between different parts of the Union. Every local Communist Party spoke of the reform of the Union, but many wished to go further, not only in encouraging republics to secede from the USSR, but even to abolish the Communist Party itself. Ukrainian society is in a state of decay that at times has approached anarchy. The party, it is sometimes asserted, will not step down and relinquish its power, yet on the other hand it has lost its ability to govern. Thus a void could develop at the top.

My prognosis therefore is somewhat pessimistic, yet that does not mean that Ukraine will not survive or even that there will be riots in the streets. The fact that there is a public opinion, the growth of Ukrainian culture, of schools and literature, the international exchanges, the growth of a multiparty system, economic sovereignty—all these can be regarded as healthy developments. There is hope too in the reaction of the coal miners from July 1989 to the present, a group which spontaneously developed organizational capacities, proving with ridiculous ease that they could even govern and administrate if necessary. Such initiative, in the face of what had happened in the past, was remarkable and augurs well for Ukraine's future. We have come to expect much from the West Ukrainian Piedmont, from the Draches, Pavlychkos, Chornovils and Shcherbaks. It is doubtful, however, whether anyone could have anticipated that the "lowly" coal miners would act as they did. But it was very significant in that it demonstrated that Ukrainians from all walks of life had taken up the struggle for—for want of a better word, perhaps— perestroika.

Most of the information provided is based on Soviet sources, culled from several visits to the Soviet Union in 1987–89, from Radio Liberty in Munich, the Canadian Institute of Ukrainian Studies in Edmonton, and through personal subscriptions to a great variety of Ukrainian and Soviet journals, magazines and newspapers. The latter may well have benefited most from glasnost. Ironically, in the case of Chernobyl, the Soviet authorities were very quick to blame Western journalists for sensationalizing the issues. But it has been only through diligent Soviet journalists that we have gained our current knowledge of the situation, and this same diligence eventually prompted the Soviet health authorities to release more information. Newspaper accounts have also provided the current picture of the environmental situation in Ukraine. It is often asked whether more attention is paid to pollution in the Soviet Union than in the West, but it seems that no matter what our current dilemmas are—and they are serious enough—the environment in Ukraine at least is at a crisis level.

1 | ECONOMIC OVERVIEW

THE UKRAINIAN ECONOMY HAS long been regarded as the backbone of the Soviet industrial empire. In addition to its well known importance to the Soviet agricultural economy, the republic's natural resources, particularly iron ore and coal, ensured that it would play a key role in Stalin's plans for industrialization in the 1930s. It was in Ukraine, in 1935, that the Donbass miner Aleksei Stakhanov achieved the improbable feat of hewing 102 tons of coal in a single shift, thereby helping to initiate the official movement named after him, "Stakhanovism" (see Chapter Six). In many ways, Stakhanovism is symptomatic of many of the problems facing the economy today: the ruthless and quasi-colonial exploitation of nonrenewable resources for the all-Union economy; the emphasis upon quantity rather than quality; the selfless performance of the workers and their apparent willingness to make sacrifices for the good of the country. Past industrialization has signified that at the beginning of the 1990s decade, when Ukraine starts on the road of economic sovereignty (and perhaps political sovereignty), it is facing the ultimate paradox that upon gaining control over the nation's natural resources, it lacks sufficient resources to survive alone.

In the Gorbachev period, the chief difficulty in implementing economic reform in Ukraine is that since the late 1970s, many of its chief industries have not been cost effective. The coal industry, as will be demonstrated in a later chapter, has been operating at a loss for some time, and with a staggering human injury toll. The production of refined steel has stagnated. The prospect-

1

ing for new ores has suggested that Ukraine's days as a key supplier of raw materials are numbered. In addition, the workforce is facing apathy, discontent and the fact that it has long contained an excessive amount of unskilled labor. All the above has been the result of centralized planning. Prior to economic sovereignty, the republic controlled only 5 percent of its own resources and industries.

Another problem was raised in the spring of 1989 by S. Dorohuntsov, Chairman of the Council for the Study of the Productive Forces of Ukraine, affiliated to the Ukrainian Academy of Sciences. He pointed out that although the republic accounts for a disproportionately high total of the gross national products of the USSR—it provides more than 17 percent of Soviet national income—the profits accrued to the state are not returned to Ukraine in the same proportions. Dorohuntsov is an advocate of regional economics, and believes that Ukraine must have the authority to alter significantly the unbalanced nature of its industrial development. He pointed out that 72 percent of gross industrial production in Ukraine up to 1988 was devoted to industrial goods, and only 28 percent to consumer items. He described the attitude to the former as: "production for the sake of production itself."[1]

Ironically in view of current economic difficulties, Ukraine began the "age of perestroika" with something of a reputation in the sphere of economic reform, largely thanks to a single enterprise, the giant machine-tool factory called "Frunze" at Sumy, in northern Ukraine. This firm produced machinery for the chemical, oil and nuclear power industries. The factory (or "association") was placed under conditions of what was termed a "deepened economic experiment" in January 1985, a sequel to the Andropov reform of 1983. In addition, the deputy director of the Sumy association, Volodymyr Moskalenko, requested that "self-financing" be introduced there. This signified that the association should pay for all of its expenses out of its own profits, an idea that evidently originated (in Soviet parlance at least) in the 1960s. Out of every ruble of profit made, thirty kopecks went into the state budget, while seventy kopecks remained at the enterprise "predominantly for additional payments to workers." The implication was that control over profits raised workers' in-

1. *Robitnycha hazeta*, April 19, 1989.

terests in their jobs by permitting them "maximum independence."[2]

Following the practices at the "Frunze" works, the Politburo
of the Central Committee of the Communist party of the Soviet
Union (CC CPSU) issued a decree in August 1985, which stated
that selected enterprises throughout the USSR would be permitted to use their own profits to update equipment and, during the
1986–90 period, enterprise funds were to be used to build residences and recreational facilities for workers. The "Frunze" was
a model enterprise and hardly representative of the average
Ukrainian firm. But one must acknowledge its symbolic importance in the economic reform, which led directly to a new law,
promulgated in June 1987 called the Law of the State Enterprise.
This law embodied the above features and added others such as
the requirement to elect directors of firms from below, and with
more than one candidate standing for office.

The "economic experiment," however, belongs to an earlier
era of economic reform. One of the characteristics of Ukraine's
economy in the period since 1985 has been its almost total failure to adjust to the new conditions of life, such as self-
accounting and self-financing at the factory level. It possesses the
oldest industrial region of the Soviet Union. Factories and steelworks, especially in Eastern Ukraine, are technically outdated
and environmentally dangerous, releasing large quantities of
toxic byproducts. To bring the situation up to date, we will survey the economic predicament of Ukraine at the start of the
1990s, using three examples: the agricultural situation, as discussed at the Communist Party Plenum of late 1988; an analysis
of the general economic situation in the republic; and a look at
two fundamental problems that have brought Ukraine to the
verge of a crisis: malnutrition and poverty among a broad sector
of the population.

THE 1988 PLENUM AND AGRICULTURAL PROBLEMS

In October 1988, a plenum of the CC CPSU was held in Kiev.
While the nature of its resolutions and discussions followed fairly

2. See, for example, Radio Kiev, March 25, 1985; R.O. Bilousov, "Ekonomi-
   chnyi eksperiment v promyslovosti," *Komunist Ukrainy*, No. 2, February
   1985, pp. 52–60.

closely those of the CPSU Central Committee plenum of September 1988, it was made clear that the problem of food shortages, particularly meat and dairy products, is especially acute in Ukraine. The picture in fact had become a very gloomy one.

The plenum's review of the economy began with a lengthy speech by the longtime party leader, Volodymyr Shcherbytsky (who was retired in September 1989, and died the following January), who noted the difficulties of what he called "the period of transition" to the new economic methods associated with self-accounting and self-financing of enterprises. Placing the blame squarely on the shoulders of the then chairman of the Ukrainian Council of Ministers, Vitalii Masol, and his colleagues, Shcherbytsky pointed out that at least one-third of Ukraine's enterprises could not be transferred to self-financing at present because their profit margin was not wide enough. Many enterprises in the field were in fact operating at a loss, including several under the Ministries of Transport, Construction and the State Agroindustrial Committee. The machine-building complex as a whole was said to be well behind in its contractual deliveries of goods.[3] (One of the authorities' avowed goals in industry has been to achieve self-sufficiency in the production of basic construction materials.)

While the quality of manufactured goods in Ukraine remained poor, the transfer of enterprises to new economic conditions saw what were described as unwarranted pay rises that were not matched by similar rises in labor productivity—a familiar problem during the transition to a more market-oriented economy. And yet economic reform in the republic was well behind schedule. Although 1989 was to have been the completion date for economic reorganization, less than 50 percent of Ukrainian enterprises had started to organize their own finances and accounting by the fall of 1988.

The main focus of this plenum, however, was not industry, but agriculture. Shcherbytsky expressed his concern about the declining per capita meat consumption in urban areas. Special commissions had been created at the central, oblast, city and raion levels, with the task of improving the supply of food to the populace. At the central level, the commission was headed by Masol,

3. *Radyans'ka Ukraina*, October 11, 1988.

which is an indication of how seriously the authorities viewed the situation.

Yet this problem was hardly new. It was noted at the plenum, for example, that during the period of the Eleventh Five-Year Plan (1981–85), the annual production of sugar beets, oil products, and potatoes declined compared with the previous plan period, while the rate of increase for food products such as meat, milk, eggs, and vegetables fell considerably. Altogether, according to a spokesman for the Ukrainian "agroindustrial complex," the rate of increase in the Eleventh Five-Year Plan was only half that of the Ninth Five-Year Plan (1971–75).[4] Because of the shortage of meat and milk products, annual consumption of foodstuffs per capita had now fallen below recommended norms (and, as will be seen below, it has continued to fall).[5]

Shcherbytsky made reference to the especially critical situation in Vynnytsya Oblast, which had failed to fulfill plans for the production of meat, milk, and eggs for the past two years. The yield of milk per cow in the oblast was reported to be sixty kilograms lower than the republican average, while the consumption of meat per resident over the past year was 6.4 kilograms less than in 1983. Vynnytsya was also said to be in a lamentable state in other respects: two-thirds of its villages did not have bathhouses, some 160 did not have hard-surface roads, and less than 1 percent were connected to the local gas network. Shcherbytsky pinpointed the first secretary of the oblast party committee, L.L. Kryvoruchko, as the chief culprit for this woeful state of affairs. Yet the lack of such rural facilities is common throughout Ukraine.[6] Shcherbytsky drew the plenum's attention to the fact that commercial outlets in the villages were also lacking in basic supplies. Many stores, he stated, had no flour, milk, oil products, fish, or even salt for sale. Another problem was the nondelivery of promised quotas of potatoes from the fields, because there was a shortfall of 27 percent of the expected supply.

In a speech that was made on the second day of the plenum, Masol provided more details about the state of livestock raising

4. *Pid praporom leninizmu*, No. 18, 1988, p. 33.
5. *Radyans'ka Ukraina*, October 12, 1988.
6. The summer 1990 session of the new Ukrainian Supreme Soviet devoted considerable attention to this same topic. See, for example, the account in *Radyans'ka Ukraina*, May 17, 1990, about the proposed renewal of small villages.

in the republic, and about cattle in particular. He revealed that in various oblasts, the number of cattle had begun to decline in 1987–88 at a rate of 5,000–8,000 heads each in Vynnytsya, Donetsk and Cherkasy oblasts. The worst situation of all was said to be in Odessa and Dnipropetrovsk oblasts, where the number of heads of cattle had fallen over the previous five years by 19 percent and 13 percent, respectively. A major reason for this development, added Masol, had been the difficulty in providing feed crops for these animals.

In the view of the Ukrainian authorities, there were several remedial steps that had to be taken immediately: better use had to be made of the cultivation of private plots and holdings, and of the leasing opportunities offered by the September plenum of the CPSU Central Committee. Masol pointed out that almost nine million Ukrainian households had private plots, with a total of about 2.5 million hectares of land. Until recently, however, these plots were being constantly reduced in size, and the private sector's role in agriculture was a declining one. A decree issued by the Central Committee plenum in Moscow in September 1988 had declared that there must be a radical restructuring of economic relations in villages, while the government organs, especially at the oblast level, had to take more initiative on rural questions.

On the whole, the discussion in Kiev was somewhat franker than in the past. For the first time, it seemed, the concern of the party leadership for the rural predicament appeared to be genuine, although it hardly dealt with the problems in great depth. Little mention was made, for example, of the arid weather conditions in the republic, although the oblasts scolded for their poor performances included Crimea and Odessa, which both suffer regularly from water shortages. It was decided that the solution to Ukraine's agricultural difficulties lay in the following:

1. economic reorganization;
2. the formation of commissions at various levels to scrutinize the food question;
3. making better use of private plots, which were known to be more productive per capita than collective and state farms;

4. expanding the development of cooperative societies, which in the fall of 1988 employed but a small proportion of the Ukrainian rural population;
5. increasing the amount of land held on lease;
6. considerably improving the selection and breeding of livestock, especially cattle.

In its general discussion, the Ukrainian Plenum explored a wide range of topics. Examining the state of the Ukrainian economy as a whole, it confirmed the demise of many once vital areas of Ukrainian industry. Thus the First Secretary of Donetsk Oblast (subsequently replaced), A.Ya. Vinnyk, noted that although the oblast's coal industry was being transferred to conditions of full self-accounting and self-financing, there was little incentive for the miners to take a greater interest in their work when no new mines were being constructed. He felt that there should be greater capital investment into the industry. Moreover, he said, even the new conditions did not always result in a better deal for the work force since, in many of the enterprises transferred to self-financing, little of the profit remained once payments had been made to the state budget.[7]

In the construction industry, headed by V.P. Salo, which had been the target of regular attacks in the press, many leaders were said to be quite unprepared to work under the new economic conditions, which provided more scope for initiative. The main problem in this sphere was said to be the long period of time required for the completion of construction projects. Salo stated that better planning was required, and revealed that thirteen planning institutes had been created for this purpose. Nevertheless, the industry was being constantly hampered by its poor supply system. In the ferrous metallurgy industry, now under all-Union rather than republican jurisdiction, a serious impediment to progress remained the overusage of metal and the continued use of obsolete equipment. In both these important industries, therefore, perennial problems had not been resolved despite the heralding of a "new era" in Soviet industry.

7. *Radyans'ka Ukraina*, October 13, 1988.

THE OUTLOOK IN 1990

In the period just prior to the introduction of economic sovereignty in Ukraine, the outlook appeared to be increasingly bleak. Ukrainian economists were stressing the need for a convertible ruble and the importance of holding down spiralling wage rises. At the same time, the republic was faced with a high budget deficit, shortages of basic foodstuffs and consumer goods, an uncertain mood among the labor force, and even the prospect of power cuts. The costs of eliminating the effects of the Chernobyl disaster were among the largest single impediments to any economic recovery, and expenses in this sphere were continuing to rise as a result of the major construction work required to house and provide other facilities for present and future evacuees. Moreover, the measures being introduced to alleviate the current economic predicament were at best half-hearted and at worst purely cosmetic.

In December 1989, *Radyanska Ukraina* published a report on the state of the Ukrainian economy by V.P. Fokin, the Chairman of the Ukrainian State Committee for Statistics.[8] Fokin noted that in 1989 the living standards of the population fell as a result of hidden price rises. At the same time, wages rose twice as fast as national income. The consumer goods industry remained the key problem area; consequently, it was planned to increase production in this sector by 5.3 percent in 1990 compared with a planned rise of only 3 percent in the output of producer goods. About 320,000 manual laborers were to be released from their jobs. Several producer-goods enterprises were being reprofiled in an effort to raise production of commodities such as footwear, knitwear, and fabrics, which were in short supply.

Fokin emphasized the desire of planners to put the financial and economic situation of Ukraine onto a normal footing. In his opinion, financial resources could be increased by reducing production costs and cutting down the number of administrative personnel in various sectors. Such measures did not, however, address the very basic problem of the structure of Ukrainian industry. Most of the major sectors of Ukrainian industry had remained under all-Union control during the Gorbachev period,

8. *Radyans'ka Ukraina,* December 2, 1989.

and critics had pointed out that this had led to some glaring anomalies. The price paid for one ton of sugar beets in Belorussia, for example, was double that paid in Ukraine. Similarly, in 1987, the average wage in Ukraine, which has the largest concentration of heavy industry in the Soviet Union, was considerably lower than in Belorussia, the RSFSR and the Baltic republics.[9] Meanwhile, Ukrainians had become short of food, basic commodities, and, above all, housing. Some 2.4 million families—or about 14 percent of the total population of Ukraine—were said to be in "extreme" need of housing. As part of the overall plan known as "Housing 2000," according to which every Soviet family is supposed to be provided with an apartment by the end of the century, a total of 105.1 million square meters of housing was scheduled to be built in 1990. This was over 17.5 million square meters more than the total amount constructed during the Twelfth Five-Year Plan. Yet even if this target was achieved (it was unclear at the time of writing whether this would be the case), the housing shortage would not be eradicated.

Traditional Ukrainian industries were clearly in trouble by 1990. Coal output, for example, had fallen to 179 million metric tons in 1989, the lowest total for several decades. Although a gradual reduction in coal output had been anticipated and this trend was expected to continue over the next few years, the drop in 1989 was much larger than predicted—9.4 million tons below the expected output for the year. The alleged reason behind the poor showing was not the coal strike of July 1989, but rather the fact that Ukrainian coal mines were now being worked only six days a week instead of seven. The steel industry was also in difficulties, reporting a considerable lagging in the reconstruction and technical retooling of enterprises. In addition, the ferrous metallurgy industry had failed to achieve the technical progress anticipated. Thus the industry in Ukraine remained geared toward open-hearth convertors, which were not only inefficient, but also responsible for the pollution of major steel centers such as Dnipropetrovsk.

The output of electric power in the republic in 1989 rose to 315 billion kilowatt hours as compared with 301 billion in 1988.[10] In 1990, output was not expected to rise at all. Two new

9. *Pravda Ukrainy,* January 5, 1990.

nuclear stations were scheduled to be added to the grid system in this year: Zaporizhzhya-6 and Khmelnytsky-2, both VVER-1000 (water-pressurized) reactors. Fokin pointed out that the abandonment of the Crimean station and unit four of the South Ukraine plant in 1989 were indicative of the public's hostility toward nuclear power. Thus electricity production had to be raised at thermal power stations, which resulted in the expenditure of some 5.6 million more tons of conventional fuel than originally planned. Because of these problems, strict limits were placed on electricity use and, in certain areas, power cuts occurred in 1990.

The consequences of Chernobyl are dealt with in depth in this volume. In terms of economic problems, suffice it to say here that some 3.5 million hectares of agricultural land and 1.5 million hectares of forest—about 12 percent of the territory of the republic—had been decontaminated by the start of 1990. Over 1,000 towns and villages had been affected by radiation. More than 100 new villages had been built for evacuees, in addition to the new city of Slavutych for plant operatives. In 1989, agricultural work was restricted or prohibited in an area covering some 500,000 hectares. Large sums had also been spent on the manufacture of 13,000 individual Geiger counters and other radiation-measuring equipment, but these remained in short supply. Areas that were not to be evacuated but were inside the general zone of control had to be supplied with uncontaminated food. In the period 1987–89, about 13,500 tons of meat products and almost 80,000 tons of dairy products were assigned from the state plan for the inhabitants of the contaminated areas of Kiev, Zhytomyr and Chernihiv oblasts. But massive costs lay ahead. A sum of five billion rubles of capital investment, three billion of which was to be spent in 1991–95, had been allocated simply to finance decontamination work already slated, new evacuations, medical services and the provision of clean food. Chernobyl thus had contributed more than any other single factor to a budget deficit in Ukraine of six billion rubles.

There appeared to be a dearth of original ideas in the economic plan for 1990. "Economizing on natural resources" and "raising labor productivity" had become familiar and practically

10. *Robitnycha hazeta*, November 10, 1989.

meaningless phrases. Self-financing and self-accounting were introduced officially into the Ukrainian economy in 1989 without appreciable effect. Regional discontent had become evident. The Council for the Study of Productive Forces in Ukraine, affiliated to the Ukrainian Academy of Sciences, had long complained of the injustices done to the republic in view of its past and present contribution to the Soviet economy. It had been maintained, for example, that "the dictatorship" of all-Union ministries led to a situation in which heavy industry was concentrated in large cities, with negative social and economic consequences, while smaller and medium-sized towns have experienced a low rate of development. Centralization, it was argued, had reduced the economic rights of Ukraine and lowered the standard of living of its people.[11]

One Ukrainian economist has maintained that the "federal administrative-command" system of running the economy led to incompetent leadership and a lack of incentives in the workplace. He proposed two steps to remedy the situation. First, market relations should be developed, and second, the effectiveness of leadership should be enhanced by giving economic independence to the republic, its oblasts, cities and raions. He wondered why a republic that was so well endowed with natural resources could be facing a high budget deficit. Did this not indicate that the old system was a failure? The vast majority of the profitable industries in the republic were and had long been under all-Union rather than Ukrainian jurisdiction. But before analyzing the question whether Ukraine could survive on its own economically, let us turn briefly to the social problems that have resulted from the economic malaise, and not least the problems of poverty and malnutrition.

POVERTY AND MALNUTRITION

The most detailed study of the decline in living standards among the population of Ukraine was published in the republic's main economic journal early in 1990.[12] The author, economist O.

11. *Robitnycha hazeta*, April 18, 1989.
12. O. Moskvin, "Analiz tendentsii zminy rivnya zhyttya naselennya URSR," *Ekonomika Radyans'koi Ukrainy*, No. 2, 1990, pp. 13–21.

Moskvin, revealed that Ukraine had a substantial impoverished sector whose wages or pensions had failed to keep pace with rising prices. Furthermore, the more affluent had been obliged to put a good deal of their money into savings because of the shortage of consumer goods. All were being affected by the shortages of basic foods, especially meat and meat products. While Moskvin's study examined the period 1986–88, his figures can also be updated with reference to others published for Ukrainian economic performance up to the spring of 1990, which indicated that the situation had actually deteriorated with time.

Moskvin noted that the current difficulties in the country as a whole were partly attributable to circumstances over which the planners had little control: Chernobyl, the earthquake in Armenia, and the sharp reduction in world oil prices. Ukrainian wages, however, had continued to rise at an alarming rate. Those of workers and employees increased by 12.7 percent during the period 1986–88, while those of collective farmers were up by 23.5 percent. In 1990, the average monthly pay of Ukrainian workers and employees—which stood at 228 rubles—was 7 percent higher than it was in 1989, and collective farmers received an average of 187 rubles per month, a rise of 15.4 percent over 1989.[13] Moskvin stated, however, that production and labor productivity had fallen well behind wage increases and also that, for many families, pay raises had not kept pace with the rising living costs.

It was pointed out that, despite pension increases, 40 percent of pensioners in Ukraine were receiving less than sixty rubles per month in the period under examination (in 1989, the USSR Supreme Soviet decreed that the minimum pension in the country should be seventy rubles per month). More than nineteen million Ukrainians (37.5 percent of the total population of the republic) were living in impoverished circumstances, with average monthly earnings per family member of 75–125 rubles (1988 figures). A further four million people received even lower incomes—under seventy-five rubles per family member. At the other end of the scale, 2.5 million residents of Ukraine (5 percent of the population) earned more than 250 rubles monthly. There thus appeared

13. *Radyans'ka Ukraina,* April 27, 1990.

to be three distinct sectors in Ukrainian society: the relatively well off, who lacked goods on which to spend their earnings; a middle stratum, encompassing about 31 percent of the population; and a larger stratum of the poor, whose standard of living was deteriorating constantly.

The first group was responsible for a rapid rise in savings deposits, which had climbed to record heights in 1986–88. Along with cash in hand, these deposits by far exceeded the level of annual commodity circulation. Over this period, state retail prices rose by 3 percent. A 10-percent rise in the price of food products included an increase of 19 percent in the price of potatoes and vegetables—evidently the only plentiful food resource, other than bread, for many Ukrainians—and a massive rise of 59 percent in that of vodka. Moskvin commented that the price rises for consumer products were not justified by improvements in quality and were partly a result of "temporarily agreed prices" for handicrafts or for fashionable goods produced by cooperatives and partly a result of speculation in imported or rare products and in goods that were in short supply. One result of the rise in retail prices was increasing inflation. An inflation figure of 2.2 percent was cited, but more recent statistics put the rate of inflation in spring 1990 in the republic at closer to 6 percent (and even this total appeared to be a significant underestimate).

The gravity of the situation was underscored by a comparison between the actual levels of food consumption in Ukraine with officially recommended norms. In 1988, the annual consumption of meat and meat products per head of population was sixty-eight kilograms, compared to a recommended norm of eighty-three. (In this context, it should be noted that Soviet figures usually include fat and secondary products that would not be considered meat by Western standards.) For fish products, the corresponding figures were 18.7 kilograms compared to a recommended twenty kilograms per head of population; and for fruits, fifty-six against a recommended ninety. Bread and grain products were relatively plentiful, as was sugar, although in the latter case, the total output figure was misleading because much of the sugar produced went into the illegal distilling of alcohol. An average citizen who went into a restaurant in the hope of meeting his nutritional requirements was unlikely to receive much satisfaction, Moskvin observed. Restaurants, cafeterias, bars and cooperatives in Ukraine were notable for their expensive and poor quality food.

As for alcohol, the campaign against it launched in 1985 by Mikhail Gorbachev went badly astray in Ukraine. Consumption of state-produced alcohol was reduced to 3.2 liters per head per annum in 1988, compared with a total of 6.3 liters in 1980. At the same time, however, the production of *samogon* (moonshine) had risen, as had the consumption of far more dangerous perfumes and methylated spirits, with the result that total consumption of alcohol had almost certainly remained at the same level (it may even have risen). The widespread use of sugar for the production of *samogon* necessitated the rationing of sugar in Ukraine after 1988.

Turning to consumer items produced by light industry, Moskvin stated that almost all Ukrainian families possessed television sets and refrigerators and that about one-third owned cars or motorcycles. The population was far from content, however. Along with the rationing of sugar and meat products, Ukrainian families had to cope with a lack of products such as washing powder and soap. Moskvin's explanation for some of the shortages was that, as a result of inadequate elucidation in the press of a projected price rise in 1991, many residents of Ukraine had bought and hoarded products as soon as they appeared in the stores. Subsequently there was an identical reaction to rumors of a rise in the price of such goods as tobacco and jewelry. In early summer 1990, as part of the all-Union Abalkin-Petrakov reform, imminent price rises were announced along with restrictions on purchases of goods by residents from outside major cities. This again resulted in mass hoarding of goods, but the "reform" was delayed by President Gorbachev following widespread public protests.

Returning to the study by Moskvin, he lamented the opportunism of "resourceful and crafty people" who were making large sums of money by exploiting shortages of goods. The economic reform, he pointed out, had not allowed for such a development; belatedly, additional taxation on high incomes had been extended to those with monthly earnings of over 700 rubles in order to offset the activities of those with "semilabor and non-labor incomes." The brunt of Ukraine's economic problems, he continued, was being borne by large families with low wages, single mothers, pensioners, invalids, pieceworkers, and families with fixed incomes. On the basis of the figures cited by Moskvin, it could be deduced that more than 45 percent of Ukrainians were living below the poverty line of 125 rubles per family member in monthly earnings. This socially dangerous situation, the author commented, was being exploited politically by nationalist elements not only in Ukraine but throughout the Soviet Union.

Figures released in April 1990 demonstrated the continuing decline of the Ukrainian economy and the living standards of the population. If real and hidden inflation at that time are taken into account,[14] then it is evident that Ukraine was experiencing a negative growth rate. Industrial discipline was said to be worsening, and meat production in the first three months of 1990 was significantly lower than during the first quarter of 1989. Few easy remedies had presented themselves for this predicament. A revised draft law for economic sovereignty had been printed in the press, but it was limited to a dry and somewhat confusing description of the expected future relationship of Ukraine with the rest of the Soviet Union that left many questions about control of industries unanswered.[15] Although the law stated that Ukraine was to have control over all resources on its territory, it was unclear how the all-Union ministries would recoup their original investment into key industrial enterprises. More militant Ukrainians—members of the Popular Movement for Perestroika (Rukh), for example—would argue that no such compensation should be paid in the light of a quasi-colonial exploitation of Ukraine by the center for the past seven decades.

Finally, while the prospect of an economically independent Ukraine (and even political independence) appeared to many to be tantalizingly close, there was at the time of writing great uncertainty as to whether the republic might first suffer some form

---

14. For a general picture of the inflation rate in the late spring of 1990, see John Tedstrom, "First Quarter Economic Results: Can It Get Any Worse?" *Report on the USSR*, Vol. 2, No. 22, June 1, 1990, pp. 5–6.
15. *Radyans'ka Ukraina*, April 29, 1990.

of economic collapse. Ukraine's natural resources have been used so unsparingly and often so thoughtlessly that some of the more traditional industries were in decline by 1990. The proportion of Soviet coal produced in the Donetsk Basin, for example, declined from more than 50 percent in the early postwar years to around 23 percent in 1989. Much will depend on what sort of economic relationship can be reached with the Russian republic that has emerged under the presidency of Boris Yeltsin, who soon made it known that his republic might wish in the future to be paid in hard currency for its commodities. The short-term future will be very difficult for the average Ukrainian citizen, but, as will become evident, the economic and ecological problems of the republic have politicized an entire nation.

2 | THE EFFECTS OF THE CHERNOBYL DISASTER

THE CHERNOBYL DISASTER HAS become one of the most divisive and controversial topics in Ukrainian society. Although it took place in April 1986, only one year after Mikhail Gorbachev became General Secretary of the CPSU Central Committee, its effects have been far-reaching. The first medical results of the accident began to manifest themselves in 1989. At the same time, commencing in 1987, republican citizens have questioned the official interpretations of Chernobyl. They have asked whether the true story about the aftereffects of the tragedy has yet to be told. In particular, many of the statements provided by the USSR and Ukrainian Ministries of Health Protection have been subjected to close scrutiny; as have those of the chief agency engaged in monitoring the victims of high radiation background, the Center for Radiation Medicine, affiliated with the USSR Academy of Medical Sciences and based in Kiev.

It is fair to say that the initial impact of the disaster has been consistently underestimated by both Soviet health and nuclear power officials. Several individuals have been singled out by critics for ostensibly misleading the public, particularly for making premature statements that there would be virtually no adverse health consequences from radiation exposure for the population of Ukraine (or Belorussia). Chief among these spokespersons have been the following: Leonid Ilyin, the Vice-President of the USSR Academy of Medical Sciences; Anatolii Romanenko, who until November 1989 was the Ukrainian Minister of Health Protection; his successor in this post, Yurii Spizhenko; Borys Shcherbyna, one of the rotating chairmen of the Government Commission that was established to eliminate the consequences of Chernobyl; and several scientists at the Center for Radiation Medicine.

The forced retirement of Romanenko came at a time when the CC CPSU Politburo was holding a debate on the current problems that have arisen from Chernobyl. He has been widely criticized for his initial failure to issue a health warning to the Ukrainian population before May 5, 1986, nine days after the event. As the Director of the Center for Radiation Medicine, he had been notably reticent about publishing or even releasing data on the health effects of the accident, and had instead repeated regularly that there have been no discernible anomalies among either evacuees or newborn babies that can be attributed to radiation exposure.[1] Nor have Soviet experts attempted to anticipate the number of likely illnesses or casualties that might result from the accident in future years. Those in the West who have done so, like the US doctor Robert Gale, have been roundly chided for their efforts.[2]

A similar picture has been forthcoming from the deliberation of US scientists on the likely effects that will ensue from Chernobyl, however. Thus it was reported that new data received have convinced experts from the Livermore Laboratory, the Institute of Electrical Energy Research in Palo Alto, and scientists from the University of California that there will be no significant changes in the state of health resulting from the explosion either of the population of the Soviet Union, or that of other countries. Interviewed by telephone for a Soviet newspaper, Robert Gale commented that although he had cited a figure of 75,000 cancer deaths worldwide as the upper limit from the accident, he had long believed that the actual figure would be on the lower end of the scale, that is, closer to zero than to 75,000.[3]

There are good reasons why Soviet and American experts should be reluctant to speculate about illnesses that may occur in the distant future. Many Ukrainians, on the other hand, have been angered at what seems to be an insensitive reaction on the part of the medical world to their plight. But not only has there

1. *Pravitel'stvennye vestnik*, No. 3, 1989.
2. David R. Marples, *The Social Impact of the Chernobyl Disaster* (1988), chapter one.
3. *Pravda Ukrainy*, January 26, 1989.

been little official light shed on medical problems, but the "official" story of Chernobyl and its consequences has been changed by the Ukrainian party authorities, anxious to show themselves in a more favorable light than has hitherto been the case. Two examples of the official reaction to a growing discontent around Chernobyl will suffice: the first from a high-ranking member of the Ukrainian party leadership, and the second from a scientist, who played the key role in the initial analysis of Chernobyl.

In early February 1989, the Ukrainian party newspaper, *Radyanska Ukraina,* conducted an interview with Borys Kachura, Secretary of the Central Committee and a member of the Politburo of the Communist Party of Ukraine.[4] Kachura tried to ridicule what he claimed were "emotional" reactions to Chernobyl while also introducing a new theme: the important role of the Ukrainian party hierarchy and the Politburo itself in the cleanup operation. This new official line maintained not merely that as a disaster the main effects of Chernobyl had been contained, but that they had been dealt with first and foremost by the Ukrainian party apparatus. Kachura's goal appeared to have been to avert once and for all any political fallout from Chernobyl on the republican party leadership.

According to Kachura, "all the republican leaders worked directly in the disaster area," and he asserted that this "fact" was well known to those working in the region. The interviewer then wondered why nothing about this activity had hitherto been revealed, noting that reports from the zone had highlighted only the bravery of pilots, miners, builders and soldiers, that is, those dealing directly with the accident's consequences. In reply, Kachura emphasized that the Ukrainian party and government had taken an active role throughout. However, in his words, the Ukrainian leaders "did not have the time to grant interviews," because they were dealing with problems as they arose. In reality, when Ukrainian party leaders had been to Chernobyl, the fact had always been well publicized. Volodymyr Shcherbytsky, the party leader until September 1989, had made only occasional visits to the designated thirty-kilometer zone around the reactor, whereas his Belorussian counterpart, Nikolai Slyunkov, was much more actively involved. So this belated attempt to amend

4. *Radyans'ka Ukraina,* February 1, 1989.

the story on the part of a Ukrainian Politburo official seems to have been pointless.

The remainder of the interview with Kachura produced some equally dubious information. He claimed that no changes to the state of health of people will result from Chernobyl, either inside or outside the Soviet Union. He praised both nuclear power specialists and the Ukrainian Ministry of Health Protection and professed to be surprised that Ukrainians were still worried about the health effects of the disaster, when science had reached more reassuring verdicts. He even defended the holding of the 1986 May-Day parade in Kiev five days after the accident on the grounds that radiation levels in the city on April 30 did not exceed the natural radiation background. When the schoolage children were taken out of the city in mid-May, this was against the advice of medical experts, who were subsequently proved to have been "correct" in making this deduction.

Other Soviet sources suggest that Kachura was mistaken with regard to the radiation situation in Kiev. The doctor and chairman of the ecological association, Yurii Shcherbak, for example, states that radiation levels were about 100 times above the maximum on the day of the May-Day parade, and that this situation was known to the party leaders who nevertheless proceeded with celebrations.[5] Moreover, scientific hypotheses about Chernobyl have all too rarely coincided with the reality. A number of Soviet writers and scholars have taken pains to reveal the chaotic and disorganized situation that took place in the days and months following the tragedy.[6] Perhaps more important, one of the main architects of the Soviet report to the International Atomic Energy Agency in Vienna in August 1986—often cited by those who concur with the official line about Chernobyl—appeared to have had grave doubts about the whole organization of nuclear power in the Soviet Union.

THE CASE OF VALERII LEGASOV

The Legasov story, while strictly speaking not a Ukrainian affair, is linked to our study because it forms an integral part of what

---

5. *News From Ukraine*, No. 40, September 1989.
6. One example is Shcherbak himself in a collection of interviews with eyewitnesses. Yurii Shcherbak, *Chernobyl: A Documentary Story* (1989).

has become the psychological effect of the Chernobyl disaster upon Ukrainians as a nation. Legasov was a well known figure throughout the Soviet Union and after August 1986 had become internationally known. The chasm between his public statements and what appear to have been his private thoughts was a wide one. Many of his doubts were mirrored among Ukrainians. Before the revelations about his private views were made public, however, some Soviet spokespersons, in Moscow particularly, spoke disparagingly of the fears of Kiev citizens about the effects of radioactivity. Fear of radiation rather than rational thought appeared to be prevalent, it was said. But what was the latter to be based on if not the quiet deliberations of scientists that should stand the test of time? After April 1988, there appeared to be a substantial rift not merely between scientists and the general public about the real results of the tragedy, but among the scientific community itself.

The importance and reliability of the Soviet account to the IAEA in Vienna has been partially undermined as a result of the mysterious death of the delegation leader, Academician Valerii Legasov, a former First Deputy Chairman of the Kurchatov Institute of Atomic Energy with the USSR Academy of Sciences. Legasov committed suicide on April 27, 1988, after two years of intensive work into the causes and consequences of the nuclear disaster. Following his suicide, which went unreported for three weeks in the Soviet Union, there was much speculation in the West over whether Legasov had been suffering from radiation sickness. According to Vladimir Gubarev, the science correspondent of *Pravda*, best known in the West as the author of the play about Chernobyl called *Sarcophagus*, Legasov was subjected to high levels of radioactivity at Chernobyl. However, it seems to have been not radiation that led to his death, but rather the adverse effect that the aftermath of the tragedy had on his psychological makeup.[7]

Gubarev has provided a brief biography of Legasov. It notes that the scientist was a graduate of the prestigious Mendeleev Institute's Faculty of Physicochemical Engineering. Following postgraduate studies in nuclear fuels, he obtained his doctorate at the

---

7. The following comments are based on the account in *Soviet News and Views* (USSR Embassy, Ottawa), No. 1, January 1989.

Kurchatov Institute. As Deputy Director of the All-Union Institute of Chemical Physics, Radiochemistry, and Nuclear and Plasma Technologies, Legasov was involved in examining the potential dangers in nuclear power engineering. Gubarev writes that Chernobyl represented a turning point in Legasov's life.

Thereafter, Gubarev states, Legasov sought a dramatic change that would overcome a perceived stagnation in Soviet science. In particular, he wanted to establish principles for industrial safety for the remainder of the century. However, his ideas were reportedly rebuffed and rejected at every turn by his fellow academicians. On April 26, 1988, the day before his suicide, a session of the USSR Academy of Sciences formally rejected his proposal by a vote of 129–100. One chemist is quoted as saying: "We don't want a rookie leading us by the nose." Legasov, who was fifty-three when he died, received the report of the session that same evening.

Views on Legasov among the academicians were, however, mixed. While a space scientist felt that his death was an "irreparable loss for science," a senior researcher at the Kurchatov Institute (whose name is not provided) believed that "Legasov was a typical representative of the scientific mafia whose politicking brought about the Chernobyl tragedy, thereby injuring the country more than the mafiosi who dealt in corruption." Gubarev, who was responsible for Legasov's memoirs being published posthumously in *Pravda,* clearly does not adhere to this view. He maintains that Legasov was a successful researcher who, though disillusioned with some aspects of Soviet science, believed that nuclear energy had a sound future. His death was a result, Gubarev implies, of a psychological breakdown.

Looking at the post-Chernobyl career of Legasov, a change can be discerned in his outlook and attitude to the Soviet nuclear energy industry, but it is not clear precisely when that change occurred. Shortly after the disaster, he remained convinced of the viability of the nuclear industry: "I am profoundly convinced that nuclear power stations are the pinnacle of achievement in power generation. ... The future of civilization is unthinkable without the peaceful utilization of nuclear power."[8] Moreover,

8. *Pravda,* June 2, 1986.

Legasov remained a firm spokesman for the view that nuclear power was a much safer and more preferable energy source than hydroelectric and thermal electric stations. On one occasion he even went so far as to say that nuclear power could become a stabilizing influence in world politics as a future struggle for the diminishing supply of raw materials might lead to conflicts between nations. At the same time, he pointed out that supplies of organic fuel could not last longer than the year 2100 and that their exploitation was thus short term.[9]

As head of the Soviet delegation to the International Atomic Energy Agency (IAEA) session in Vienna, in August 1986, Legasov adopted the posture of staunch defender of the continued use and expansion of nuclear power in the Soviet Union and he did not appear to lack confidence in the design of the graphite (RBMK) reactors of the type used at Chernobyl. The Soviet report on the accident, which he coauthored, appears in retrospect to have been an optimistic document. Widely praised for its openness and scientific accuracy—though it can hardly be regarded as an attempt at a definitive account—the report skimmed over or ignored altogether many of the more controversial issues, such as the delayed evacuation process, the dangerous nature of the cleanup work, and even the radioactive fallout that took place after May 6, 1986.

At some point thereafter, Legasov began to question the morality of technological expansion in the Soviet Union. Despite a statement to the contrary from his former institute, there is no question that his doubts, which were beginning to emerge, stemmed directly and perhaps even solely from the aftermath of the Chernobyl disaster. "The problem today is the proliferation of all sorts of projects and the concentration of vast power," he stated in an interview with Dr. Yurii Shcherbak, published in the summer of 1987, ostensibly referring in the latter instance to the USSR Ministry of Power and Electrification and the USSR Ministry of Nuclear Power (as it was known at that time). Initially, he believed, the decisions made in the nuclear power industry were good ones, but problems arose when these ideas were applied indiscriminately on a large scale. As he informed Shcherbak: "The

9. *Novosti*, March 17, 1987.

need for electric power is great. It was necessary quickly to introduce and master power (production) on a new scale. . . . The
number of people involved in the preparation of installations and
their running increased sharply. But the teaching and training
methods could not keep up with the pace of development."[10]

At this time, in late 1986, Legasov had begun to elaborate a
new philosophy of safety. The key to the problem, in his view,
lay in the relationship between "man and machine." The latter
had to be made reliable enough to withstand the inevitable mistakes of the former. This apparently straightforward logic actually represented a fundamental change in Soviet thinking which,
as far as industrial development was concerned, had overrated
the infallibility of the machine. "We have become too carried
away by technology," asserted Legasov. Although technology
had been created at the outset "in the spirit of the great humani-
tarian ideas," it had been taken over by "technocrats" who operate without moral principles: "The low technical level and the
low level of responsibility of these people is not a cause but an
effect. The effect of their low moral level."

Legasov and some of his colleagues tried in vain to draw attention to this problem, notes Gubarev, but were ignored by the
scientific community. They had sought, he stated, better comput-
ing power and simulators for the training of staff at nuclear
power plants, but to no avail. According to the Belorussian
writer Ales Adamovich, Legasov did not rule out the possibility
of another Chernobyl-type accident at one of the fourteen graphite reactors still operating in the country. "You can put it on record. I am convinced, unfortunately. The most important con-
tributing factors to the Chernobyl accident have not been and
cannot be removed. They include faults, resulting from poor construction and the lack of reliable emergency systems for similar
plants, and the impossibility of constructing concrete 'cones' to
seal them at this stage."[11]

At the time of his death, therefore, it seems clear that Legasov
was an unhappy and disillusioned man who felt that the concerns to which he had devoted his career after Chernobyl were
being ignored by his peers. He had begun to praise the cleanup

10. *Yunost'*, No. 7, 1987.
11. *Moscow News*, No. 29, 1988, p. 10.

workers at the site, particularly those carrying out very hazardous duties in the first days after the disaster.[12] Yet his colleagues seemed to be ignoring the lessons of Chernobyl and continuing to make plans for expanding the industry as though nothing had changed. Since a strong movement against nuclear power was emerging in the Soviet Union, his associates, in their turn, may have felt that his statements amounted to a betrayal.

Much of the mystery surrounding Legasov's death has remained. Indeed, Gubarev's article only gives rise to further questions. When did Legasov's change of opinion about Chernobyl take place? The interview with Adamovich is not dated. Was he regarded within the Kurchatov Institute and the USSR Academy of Sciences generally as an "old-style bureaucrat" or as a man determined to put into operation the principles of perestroika in the sphere of nuclear energy? Why has his searing criticism of the attitude of the so-called technocrats and the lack of safety at RBMK reactors never been refuted or discussed publicly by scientists? And were there other reasons for his death? Gubarev makes it plain that he, at least, is far from satisfied with the investigator's report into Legasov's death, which stated that: "we looked into the theory that Legasov was driven to suicide, but could find no proof. Legasov was independent financially and otherwise, and we uncovered no systematic attempts to humiliate him or his honor. We believe no one is to blame for his suicide."

Gubarev and Shcherbak evidently adhere to the theory that the suicide was in part an attempt to draw attention to the unfortunate state of affairs in the Soviet nuclear industry. Gubarev stated in the aforementioned article that Legasov's suicide was an act of courage rather than weakness. In the final analysis, however, whatever the scientist's motives and state of mind, he certainly succeeded in undermining the credibility of those involved in the Soviet nuclear sphere after Chernobyl. The psychological effect of his posthumously published statements on Ukrainians worried about the true effects of the tragedy and the nuclear power program can only be imagined. Further, it was followed by a newer and deeper inquiry into the radioactive fallout and its effects on Ukrainian and Belorussian territories in particular.

12. *Pravda*, May 20, 1988.

A SECOND LOOK AT CHERNOBYL

In 1988–89, Soviet scientists and government officials paid renewed attention to the impact of Chernobyl. In Ukraine, scholars began to reanalyze and to review the events and current situation. We will focus here on two studies by B. Kurkin and D. Grodzinsky, which are notable for both their thoroughness and the radical nature of their conclusions.

In a series of articles that were published in two Ukrainian newspapers, Kurkin, a Candidate of Juridical Sciences, in examining the pros and cons of nuclear energy, also focused on radioactive fallout. Noting that the official Soviet report to the IAEA in Vienna states that 3.5 percent of the contents of the reactor core were released into the atmosphere, he estimates that this would have constituted sixty-three kilograms of highly radioactive products. In fact, he maintains that the actual figure would have been higher than this total. The report cited radioactive releases from April 26 to May 6, 1986, whereas the releases continued until May 10, 1986. The figure can be compared with other nuclear explosions, Kurkin continued. The explosion of a one-kiloton atomic bomb would release into the atmosphere about thirty-seven grams of radioactive products.[13]

Turning to the comparison that has often been made—between Chernobyl and Hiroshima—Kurkin pointed out that the twenty-kiloton bomb dropped on the Japanese city at the end of the Second World War produced some 740 grams of radioactive substances. Thus, he concluded, the emission of radioactive products at Chernobyl exceeded that at Hiroshima by more than ninety times. However, he added, it would be more accurate to say that in terms of objective factors, the two events simply cannot be compared. Thus whereas wind had dispersed radioactive products after Hiroshima, at Chernobyl the fallout was much more heavily concentrated around the explosion site. Kurkin also draws the comparison with the 1979 nuclear mishap at the Three Mile Island plant in Pennsylvania, USA, noting that if one simply takes the amount of radioactive iodine released into the atmosphere, then Chernobyl surpassed Three Mile Island in quantity by more than three million times.

13. *Molod' Ukrainy*, February 15, 1989.

The Ukrainian biologist Dmytro Grodzinsky has also made a detailed investigation into the effects of the Chernobyl disaster on the immediate environment. Grodzinsky is a corresponding member of the Ukrainian Academy of Sciences, and the head of the section for biophysics and radiobiology with the Academy's Institute of Botany. He is also a Deputy Chairman of the ecological association *Zelenyi svit* (Green World). In a lengthy interview conducted in the summer of 1988, Grodzinsky expounded on his conclusions about the effects of Chernobyl.[14]

Grodzinsky noted first that the radioactive fallout landed in an area dominated by agricultural land, with from 10 to 40 percent of the territory occupied by forests. It affected a number of very different types of crops. The situation was made worse, he stated, by the fact that there had been no similar accidents previously. Chernobyl threw into the atmosphere more than 450 types of radionuclides. In the early post-accident days, about 80–90 percent of the radioactivity was made up of the short-lived iodine-131. Subsequently iodine was replaced by longer-living isotopes such as ruthenium and rhodium. But today, said Grodzinsky, the key problem lies with cesium-137 and strontium-90. He described the particles produced as "everlasting wanderers," that is, they cannot be absorbed by plants and are carried by the wind from place to place. These factors also render Chernobyl a completely unique disaster both from the nature of the radionuclides and the amount of territory encompassed by them.

Grodzinsky is very skeptical about the so-called threshold of significant radiation doses for the human organism. He believes that the majority of radiobiologists today adhere to the view that there is no such thing as a threshold dose, and that a very small rise in the background level of radiation can bring danger. Specialists today, he pointed out, are very poorly acquainted with the principles of protecting the organism from radiation. As a result, most radiobiologists maintain that "even a very low dose of radiation is harmful." However, after the Chernobyl accident, "some unqualified medical workers" had gone so far as to state

14. The following summary is based on *Nauka i tekhnika*, No. 12, 1988. Grodzinsky was promoted to a full member of the Ukrainian Academy of Sciences in the summer of 1990. See *Radyans'ka Ukraina*, May 25, 1990.

that low doses of radiation were not unhealthy for the human organism, as one can grow accustomed to low doses. This notion, states Grodzinsky, is totally incorrect. For a human being to become accustomed to such additional doses would require an evolutionary change.

Next, he took aim at the theories of the stochastic and non-stochastic effects of radiation. In the latter case, the theory is that with large doses of radiation, it is possible to forecast with certainty how many people will become sick or die. But the stochastic effects are less certain as they apply to low-level radiation. These latter effects are measured in terms of the "collective dose of irradiation": the sum total of irradiation of the majority of the population. Grodzinsky notes that these effects could occur in future generations, somewhat akin to a "slow-operating bomb." To estimate the likelihood of their appearance, he maintains, one has to assess the "risk coefficient."

But how is one to determine the risk coefficient? Prognoses have been based largely on the experience of those people who suffered at Hiroshima and Nagasaki, who were subjected to x-ray therapy. These data are in dispute today, states Grodzinsky, and are being subjected to revisions. In addition, there will "without doubt" have been mistakes in the calculations upon which the aforementioned prognoses were based. Commenting on the "violations of the endocrinal balance" that irradiation can bring, he affirms that among the early manifestations of its effects are simple illnesses like the common cold or nervous diseases, resulting from the general weakening of the immune system. Looking specifically at the effects of Chernobyl, he reveals that immediately after the accident in the city of Kiev, there was a surprisingly high death rate among rodents, especially rats and mice.

In theory, rats in particular might have been expected to survive the effects of radiation relatively well: witness their endurance of nuclear weapons tests at the Bikini atoll. Also, in Kiev, the level of radioactivity, while it may have risen to about one hundred times normal background, was not so high that it could have caused their deaths. Grodzinsky's explanation was that radioactive iodine produced by the explosion accumulated in the animals' thyroid glands and brought changes in their immune function, affecting the responses of the organisms. As a result, the rodents died not from radiation, in his view, but from their failure to combat epidemics that were breaking out among them. Humans are more resistant to such factors because they can use vitamins to prevent such a breakdown of the immune system. The problem after Chernobyl, however, was that many people in the Kiev Oblast had stopped eating vegetables for fear that they were contaminated.

Grodzinsky provides a careful examination of what some Soviet scientists have referred to as "radiophobia"—the fear of radiation that prompts people in the fallout zone to attribute every illness, however minor, to the effects of radiation. It is well known that after the disaster, some victims felt that red wine and vodka might be a suitable precaution against the onset of radiation sickness. Ethyl alcohol was thus considered to be a "radioprotector." In Grodzinsky's view, there is some truth in this notion. Spirit raises one's stability against radiation by 1.13 times, he observed. Unfortunately, it also destroys the molecules of vitamins that are themselves radioprotectors, so that its overall effect is negative.

Next, and perhaps most significantly, the biologist gives a thoughtful analysis of the effects of the radioactive fallout that is notably more frank than those of many other spokespersons. He points out that the collective dose of irradiation, upon which various prognostications have been made about future casualties, is not constant and static, but is changing all the time as radioactive particles are taken up into the food chain. He provided the first indication also that the area of serious radioactive fallout might be much wider than the designated thirty-kilometer zone around the damaged reactor. Thus he observed that there existed serious dangers of contaminated products entering the food chain in Narodychi Raion (Zhytomyr Oblast) and Poliske Raion (Kiev Oblast). He recommended the usage of potassium fertilizer to combat cesium uptake into the food chain, and calcium to prevent similar absorption of strontium.

Finally he looks at the special zone itself, and some strange occurrences around reactor four: there has been the sudden appearance of unusually aggressive roosters; packs of dogs that range like wolves; and curious crop mutations that appear to defy the rules of nature. Thousands of ducks have flown into the zone, accumulated radioactive substances, and then spread them over a wide region. The radioactive particles now collected in silt can be carried by flood waters into fields, a process that he says would render "a great mass of fruitful land radioactive." In conclusion, while doubts may be expressed about some individual points, Grodzinsky makes a powerful case for a revision of "expert forecasts" about the medical consequences of Chernobyl. His article is an effective rejoinder to the statements of officials who appear to have underestimated the effects of Chernobyl. At the time these remarks were published, they at once brought into question the safety of the Chernobyl zone for the future.

DEBATING THE FUTURE OF THE ZONE

Following the April 1986 accident, the Chernobyl nuclear power plant was quickly returned into service.[15] Yet a virtual dead zone remained in northern Kiev Oblast and the southern regions of the Gomel Oblast of the Belorussian republic. In October 1988, the discussion of the future of this area reached the pages of *Pravda,* which disputed an alleged statement from members of the Kombinat production association—which along with "Spetsatom" is in charge of the decontamination effort in the area of fallout—that the town of Chernobyl itself might have to be razed to the ground.[16] The newspaper suggested that the city could be decontaminated and repopulated, as had been certain cities in Belorussia that reportedly received higher levels of radioactive fallout than Chernobyl.

Ironically, the discussion about razing the city came only two weeks after an announcement that the thirty-kilometer zone had been cleansed. There had been little information hitherto to suggest that the city of Chernobyl would be abandoned. In fact, everything had suggested the contrary. Chernobyl was used as the headquarters of the massive campaign to decontaminate the zone, and workers have lived there on a shift basis, housed in the apartments left deserted at the time of the evacuation in May 1986. In a similar way, some members of the Kombinat production association, who are not permitted to live in the new city

15. The first two reactors were brought back on line by November 1986. Unit three, which had been more heavily contaminated, was returned to the grid in December 1987.
16. *Pravda,* October 8, 1988.

that has been built for nuclear plant operatives (Slavutych), had to live in the contaminated and largely deserted city of Prypyat, at great personal risk to their health.

In the Fall of 1988, new attention was paid to the question of radiation levels from Chernobyl fallout, in both Ukraine and Belorussia. In a bitter article that was published in the monthly journal *Novyi mir*, the writer Ales Adamovich questioned many of the long-held assumptions about the consequences of contamination, particularly in the southern regions of Belorussia. According to Adamovich, the real scale of the disaster had been deliberately played down so that it would not affect the future program for the construction of nuclear power stations.[17] He wrote that in Belorussia, the public had never been given adequate information about the levels of radioactive fallout: instead these levels had simply been averaged out per oblast rather than documented for individual raions. In cities such as Bragin, he noted, it was still considered dangerous for doctors to live there permanently, and so they had been replaced on a regular basis by colleagues from outside the region. Women and children, on the other hand, had remained there from the first.

Adamovich's view—an outspoken one at that time—was that the entire area around Chernobyl, and areas well beyond the official thirty-kilometer zone, were considerably more dangerous than had been revealed thus far by the Soviet authorities. Many Ukrainians were also sensitive to the threat of radiation after the summer of 1988, when the Kiev newspapers began to publish in their Friday editions the radiation levels in the city alongside the weather report.

In mid-January 1989, the debate about the future of the Chernobyl area was renewed, with an analysis by Mikhail Sedov, the general director of the Kombinat production association.[18] Sedov maintained that, contrary to rumors, there were no secret plans to destroy the city of Chernobyl. He noted that twelve older residential buildings had been pulled down, and a further forty-two out of 2,278 buildings were to follow, largely because of their radiation contamination, sanitary or fire hazards. Fire has been a particular risk because radiation can be spread quickly by the

17. *Novyi mir*, No. 9, 1988, pp. 167, 169.
18. *Molod' Ukrainy*, January 17, 1989.

smoke. The central streets of Chernobyl, however, had been thoroughly decontaminated, declared Sedov, together with residential and administrative buildings and markets. Sedov stated that the city's infrastructure had not only been restored, but considerably renovated.

This was not to say, however, that there was any possibility of former residents being able to return home in the near future. There were still entire zones, Sedov stated, where radiation levels were at ten to fifteen times the normal background. Further, the effect on people's health of long-duration radionuclides such as plutonium, strontium and cesium remained uncertain. In the area to the north of the Chernobyl plant, he stated, the mobility of these radionuclides into the natural environment had increased, and agricultural and animal products in this territory will be unfit for consumption for a number of years.

The Novosti correspondent who interviewed Sedov also raised the question of former residents who have returned "illegally" to their homes. Sedov admitted that about 1,000 persons were currently living in the zone and that they had come back "of their own free will." These people were said to be ignorant of the dangers of radiation, and even scornful about its possible effect on their health. Although the returnees were for the most part elderly, they were said to have been visited regularly by relatives, including grandchildren. Reportedly some sixty children had visited evacuated villages at various times during the summer of 1988, for example.

Prevailing conditions rendered any return to a normal pattern of life impossible, said Sedov. In the village of Opachychi, which has 103 residents, radioactive contamination on clothing and other objects was said to be two to three times the permissible level for people working on shifts at the nuclear plant. Sedov may have been referring here to the relatively high radiation norms for those engaged in cleanup work, which can be up to twenty-five rems per worker.[19] Drinking water from wells in villages such as Opachychi, Illintsi and Kupuvate showed levels of contamination that were considerably in excess of approved medical limits. And even if these villages could be decontaminated, there was the problem of secondary radiation in areas that had already been cleaned up.

19. See the Soviet account, *Chernobyl': Segodnya i zavtra* (1988), p. 41.

Sedov mentioned a meeting held in Chernobyl in October 1988, to discuss whether it was acceptable to have people living and farming in the zone. Although a number of villages have been resettled and farming resumed in various parts of the zone, the meeting concluded that resettlement was undesirable and that those who had returned should be asked to leave. In the zone around the nuclear plant, all types of activity other than research and a minimal amount of experimental farming should be prohibited. "Science has declared" that it will be impossible to live there permanently for a long time.

Even experimental use of the land produced problems since, when the soil was plowed, it was virtually impossible to prevent the spread of dust; consequently, secondary radiation had repolluted sections that had already been cleansed once. Moreover, secondary plowing of the soil tended to lead to a worsening of the situation as biochemical processes were activated. Sedov concluded by stressing the relatively open policy adopted by the authorities at Chernobyl and the availability of information about the zone as provided, for example, by the Kombinat workers' newspaper, *Trudovaya vakhta,* and for those unable to obtain it,[20] the telephone information service at the cleanup headquarters.

Despite this relative frankness, it was becoming clear to many observers that much of the story about the effects of Chernobyl remained to be told. Adamovich had issued a warning about the situation in Belorussia, and it was from this republic that there came the announcement early in 1989 that twenty villages in the Gomel and Mogilev oblasts were to be evacuated, as it had been discovered that they contained several "hotspots" of intensive radiation. The explanation offered by the Chairman of the Belorussian Government Commission established to remove the consequences of the accident in this republic, Deputy Chairman of the

---

20. This newspaper is unavailable outside the thirty-kilometer zone and cannot be exported abroad. During a visit to Chernobyl in the summer of 1989, it was possible to acquire a single copy. Ironically, one of Kombinat's biggest problems is the lack of public recognition for the work undertaken, much of which is of a highly dangerous nature. If the newspaper were permitted a wider readership such recognition might be forthcoming, for there could hardly be a better indicator of the day-to-day existence of a Chernobyl cleanup worker than its columns. Here again, however, official secrecy has had lamentable results.

Belorussian Council of Ministers, V. Yevtukh, was that glasnost had made slow progress in the republic. It transpired that in total, eleven raions of Gomel Oblast and six in Mogilev were suffering from serious cesium fallout. This territory encompassed about one-fifth of the republic, with 415 population points and 103,000 people.[21]

THE SITUATION IN SPRING 1989

Following a visit to Chernobyl by General Secretary of the CC CPSU, Mikhail Gorbachev, in February 1989, and in response to what was described as a disturbing lack of attention to the problems caused by radioactive fallout from the damaged reactor, the Information department of the Ukrainian Council of Ministers published the first detailed account of the consequences of the accident.[22] The article was notable for the provision of a map that divided the fallout region into four zones: a zone of alienation (I); a zone of temporary evacuation (II); a zone of constant control (III); and a zone of periodic control (IV). Zone I had been completely evacuated, but it contained the active Chernobyl nuclear plant, whose workers had recently been transferred from a routine of shifts to one of regular work, and were being bussed in from Slavutych,[23] Zelenyi Mys, Kiev or Chernihiv. Zone II con-

21. *Izvestiya*, February 1, 1989; *Radio Moscow*, February 9, 1989. According to the Soviet news agency TASS, the affected areas in Belorussia were divided into three zones. Close to the nuclear plant itself, 4,400 people were evacuated during the month after the accident. In the second zone, in the period from approximately May to September 1986, a further 19,000 were removed from a "second zone." The evacuations announced in February 1989 referred to a third zone, following belated analyses of 40,000 collected soil samples in the Belorussian republic.

22. *Radyans'ka Ukraina*, March 1, 1989. There were several aspects of the disaster that were addressed here for the first time. It was reported, for example, that to assist in the examination of accident victims, 274 doctors were sent to Kiev Oblast in 1988, and 352 to Zhytomyr Oblast (it is not clear where they came from). This indicated, in the view of the authors of the article, that medical cadres were being strengthened. Yet it could equally well have demonstrated that there was a growing concern over health problems in Zhytomyr Oblast (see Chapter Three).

23. Slavutych is the replacement city for Prypyat, the former town of operatives at the Chernobyl nuclear power plant. Despite a number of problems in its construction, Slavutych possessed a 1989 population of around 10,000. Originally it was anticipated that it would grow to 30,000. There have been some emotional articles both about the "loss" of Prypyat and the emergence of the "21st century city," Slavutych. On the former, see the comments by the for-

tained the estimated 1,100 people who had returned to their homes without permission, noted above. Zones III and IV embraced a total of 176 settlements with about 84,000 people. Of these, about 47,000 were said to be living in villages in which the contamination of the soil by cesium-137 exceeded fifteen curies per square kilometer (the official maximum limit for cesium fallout), and food had to be brought in from the outside. Further, Zones III and IV included territories that had not been believed to be contaminated. Several villages in the northern part of Rivne Oblast, for example, were located within them, as was a section as far south as Ivankiv (Kiev Oblast), very close to where the first evacuees were taken for "safety" after the disaster. Four Ukrainian oblasts were now revealed to have been seriously affected by the radiation: Kiev, Zhytomyr, Chernihiv and Rivne. Plainly, many residents had been living for nearly three years in ignorance of the danger that surrounded them.

According to the data provided, private farms in the four fallout zones were safer than state or collective farms, primarily because they were better manured, and because the private sector had been given a leaflet on "The Introduction of Special Private Farms on Contaminated Territories." Private plots had also received an increased amount of phosphorus and potassium to help prevent cesium from entering the food chain from the soil. But the forests of northern Ukraine were now sprinkled with warning signs because they were severely contaminated, and new rules were issued under the title "Temporary Regulations for Forestry Operations." The upper reaches of the Kiev Reservoir were also declared off limits to fishermen, which certainly had not been the case in the first year after the accident.

The levels of contamination of food in parts of Zones III and IV were alarmingly high. In 1988, for example, research into the quality of milk in the private farming sector revealed that in 30–50 percent of cases, the permissible levels of contamination were exceeded. Also in excess of these limits were 30–50 percent

---

mer construction chief of the Chernobyl station, V.T. Kizima, in *Radyans'ka Ukraina*, May 7, 1989. For an account of how the site for Slavutych came to be selected, see *Stroitel'naya gazeta*, February 19, 1989. As noted below and in Chapter Three, there are fears that Slavutych may have been located on a "radioactive patch."

of mushrooms, 10–12 percent of meat, and 20–30 percent of the fish caught in reservoirs and lakes.

Unfortunately, the health of those affected by these conditions was not discussed comprehensively by the Information department, possibly because this information emanates from parts of Soviet society that have been accused of excessive secrecy: the Center for Radiation Medicine and the USSR and Ukrainian Ministries of Health Protection. It is made clear, nonetheless, that the year 1986, in comparison to the average during the period of the Eleventh Five-Year Plan (1981–85), showed a distinct rise in anemia and diseases of the respiratory organs and the stomach in the areas under investigation.

In April 1989, Leonid Ilyin also produced some new information on the possible medical consequences of Chernobyl, though most of his statistics were restricted to the Gomel Oblast of Belorussia. From the monitoring of 786 zones affected by the disaster, Il'yin stated that in the period from April 26, 1986 to January 1,1990, the real mean radiation dose received by the people in the area under observation would be six rems (a revision of earlier higher estimates). He stated that 2,600 people would receive a high dosage of 17.3 rems, and 800 people would get more than this figure.[24] While this would signify that 3,400 people from the area under study would have been seriously affected, there was no indication of how far factors such as the lack of health warnings and basic precautions were taken into account.

Ilyin anticipated that in Gomel Oblast, the number of excess thyroid cancers over the next thirty years would be sixty, and the number of fatal cases of cancer and leukemia in this same region would be sixty-four. Yet one could turn from scientific supposition to reality. Already there was increasing evidence that children especially had been affected by radiation-related illnesses in some remote areas (see Chapter Three). Some regions had reported "an ominous deterioration in children's health" that included swollen thyroid glands, headaches and nosebleeds.[25] In short, they had incurred symptoms that could be attributed or related directly to radioactive fallout.

24. *Sovetskaya Belorossiya,* April 16, 1989; and *Vechirnii Kyiv,* April 26, 1989.
25. *Sem'ya,* No. 17, 1989, p. 3.

The situation on the third anniversary of the disaster was reviewed extensively by the Ukrainian media. On April 20, a press conference was held in Kiev, and discussed whether it might be possible to issue the population with individual Geiger counters to measure radiation levels. Andrei Pralnikov, a journalist with the weekly *Moskovskie novosti,* published an angry article which focused both on the poor quality of Soviet Geiger counters and their scarcity. He noted, for example that "even the instrument used at Chernobyl was thought highly of. On several occasions I saw people beg workers to sell or exchange their "peepers"—in the zone it became a kind of hard currency, as valuable as a field set of fatigues worn by our men in Afghanistan."[26] However, according to the First Deputy Chairman of the Ukrainian Council of Ministers, although a Kiev factory had begun to produce dosimeters, based on plans provided by the Ukrainian Academy of Sciences, the price of one is a prohibitive 450 rubles, or slightly less than three month's salary for the average rural worker.[27]

The real focus of the anniversary date was the continuing operation of the Chernobyl plant. Among the most frequent respondents to questions from the Ukrainian public were Sedov and Mikhail Umanets, the Director of the Chernobyl plant. Sedov informed readers of one Kiev newspaper that the current radiation levels at Prypyat varied from 0.2 to 2.0 millirems per hour (up to 200 times the natural background), while those at the town of Chernobyl ranged from 0.1 to 1 millirems per hour. The conclusion was that while Chernobyl may be unfit for permanent habitation for "tens of decades," Prypyat may never again be fit for residence.[28]

To the sequential question as to why members of the Kombinat production association were permitted to live in such cities, Sedov replied that they still work there on a shift basis only. The goal was to transfer eventually all the cleanup crews to Zelenyi Mys (Green Cape), the settlement for shiftworkers constructed on the Kiev Reservoir in 1986–7, which was to be expanded. "Believe me," he stated, "People do not live here because they want to." In short, the Kombinat workers were making sacrifices to their health by remaining in this risky zone.

26. *Moscow News,* April 30, 1989.
27. *Radyans'ka Ukraina,* April 23, 1989.
28. *Robitnycha hazeta,* April 26, 1989.

But the continuing operation of the station puzzled many observers and invoked anger on the anniversary date. In Kiev, a reported 12,000 people gathered in the soccer stadium to commemorate the occasion. One speaker was Dmytro Pavlychko, a poet, chairman of the Shevchenko Ukrainian Language Society and subsequently a prominent member of the Popular Movement to Promote Perestroika in Ukraine, known as the "Rukh." Pavlychko declared that all the lessons of Chernobyl had not yet been learned. Henceforth, he maintained, anyone being sent to work at Chernobyl should go there for one of two reasons: either to dismantle the station, or to assist in sanitizing the zone. Nuclear power operation must be guided by the wisdom of the people, he stated. Kachura and other Ukrainian party leaders attended this meeting and probably would not have agreed with such sentiments, but clearly Pavlychko's speech was among the most warmly received by those present.[29]

Umanets acknowledged the clamor for the shutdown of the Chernobyl plant, but said that he preferred to look at the question from a scientific viewpoint. He declared that there is a possibility that the station could last out its thirty-year lifespan, which would mean that reactor one would be decommissioned in the year 2007. However, faults with the graphite reactors (which had by now been removed from the future building program) were necessitating detailed reconstruction work every fifteen years. Therefore work was being conducted at the Soviet Union's oldest graphite plant at Leningrad to estimate whether such stations could remain cost effective. If it transpired that such reconstruction work was not viable economically, continued Umanets, then Chernobyl, along with the other graphite reactor stations, would be shut down within three years.[30]

Logically, it appeared to many that by keeping the Chernobyl plant open, the authorities were ipso facto prolonging the insecurities not only of 3,500 plant operatives who have to make the

---

29. *Radyans'ka Ukraina,* April 28, 1989; and *Robitnycha hazeta,* April 28, 1989.
30. *Izvestiya,* April 26, 1989. Umanets's comments pertained only to one problem with the graphite reactor. The ostensible reason for the cessation of building work on others lay in its inherent technical flaw: that it becomes unstable at less than 700 megawatts thermal power. The scientific "weakness" of the reactor has also been acknowledged at a Soviet nuclear society meeting. See *Stroitel'naya gazeta,* April 21, 1989.

daily journey there from Slavutych, but also of the Kombinat workers. This was of far more importance to critics than the economic questions around the graphite reactor. Also on the anniversary date, the Ministry of Power and Electrification of the USSR appeared to take official secrecy one step further with the adoption of a new law to prevent the press from publicizing accidents at nuclear power plants, even if they result in nonfatal environmental contamination.

The anniversary date elicited mixed concern among Ukrainian and Belorussian party functionaries. At the commemorative meeting in Minsk, for example, no government or party functionaries were in attendance. In Kiev, an international symposium began on April 25, with delegates from seventeen countries, including the United States and Canada, which studied the effects of conventional war in Europe, looking at factors such as the density of nuclear power plants.[31] The title of this meeting was "EuroChernobyl," but it related perhaps more to the theme enunciated by Mikhail Gorbachev of a "common European home." The meeting was warmly endorsed by activists such as Yurii Shcherbak of the Green World ecological group, but it appeared to be something of a diversion from the real problems being caused by the Chernobyl tragedy.

AN EXAMINATION OF THE THIRTY-KILOMETER ZONE

In June 1989, it was possible to visit Chernobyl on the invitation of the Ukrainian Ministry of Foreign Affairs. In addition to conducting interviews with the Kombinat production association, the Director of the Chernobyl nuclear plant, Mikhail Umanets, and the directors of the experimental hothouse run by "Kompleks" in Prypyat, it was also possible to look around the zone, take radiation measurements with a Geiger counter (supplied by Kombinat), and to examine the operation of the nuclear plant itself. The aim in so doing was to provide answers to questions arising about the zone, particularly regarding the continuing operation of the plant despite acknowledged technical faults, and whether any parts of the zone were likely to be habitable in the near future.

31. *Radyans'ka Ukraina*, April 26 and 27, 1989.

KOMBINAT PRODUCTION ASSOCIATION

Chernobyl in the summer of 1989 was a bustling town, with some 6,500 shiftworkers living there, the majority of whom appeared to be based more permanently at the shift settlement of Zelenyi Mys. At the approach to the town, military reservists—adolescents in brown overalls—were much in evidence, often sitting somewhat alarmingly in the undergrowth taking "smoke breaks," despite road signs that warned of high radiation levels in the ditches. Along with my guide, Yurii Risovanny, a leading engineer with the international department of Kombinat, I arrived at the headquarters of the Kombinat production association to be greeted by Pavel G. Pokutnyi, the Chairman of the Department of Information and Foreign Relations, with Kombinat.[32]

During our meeting, Pokutnyi explained the various units encompassed by Kombinat. At that time, it included the three operating units at Chernobyl, the specialized unit Kompleks, which dealt with decontamination work and the problem of radioactive waste, and the construction of the city of Slavutych (Chernihiv Oblast), which in mid-1989 had a population of 10,000. Altogether it comprised, under General Director Sedov, nine sections:

1. the Chernobyl nuclear power plant;
2. the Kompleks specialized unit;
3. the Radiation Monitoring department;
4. the Heat and Power Supply department;
5. the Dispatching and Process Control department;
6. the Personnel Catering department;
7. the Auto-Transport department;
8. the Housing and Communal Services department;
9. the Construction Project department (Slavutych).

Because of the advancement of work at Slavutych, the plant operatives at Chernobyl were taken off shiftwork in December 1988, and placed on a regular routine. However, although these workers travel from Slavutych by rail, the trains are changed at the entrance to the zone in order to prevent the spread of radioactivity outside the zone. Pokutnyi acknowledged that there had been a discussion in the press on whether Slavutych was built on

32. Interview with Pavel Pokutnyi, Chernobyl, Kiev Oblast, June 14, 1989.

a radioactive patch. However, he dismissed this possibility as a speculation, and declared that in terms of the natural radiation background, it had been constructed in one of the cleanest of areas.

In addition to the 9,000-strong Kombinat team, the zone was also populated by an association geared for emergencies called "Spetsatom" (which had taken on the appearance of a rival, parallel organization rather than a complementary one), and by a growing squad of scientists. Reactor unit four itself was by then under the charge of the Kurchatov Institute of Atomic Energy, directed by one of the main personalities involved in the initial decontamination campaign, Evgenii Velikhov, the Vice-President of the Soviet Academy of Sciences. Scientists from twenty-six institutes of the USSR and Ukrainian Academies of Sciences were examining the influence of radiation on the surrounding plant life. In the near future, Pokutnyi explained, a scientific and technical center was to be opened in Chernobyl, with the cooperation and participation of scientists from foreign countries. (Subsequently this Center was organized in Zelenyi Mys. At the time of the interview, non-Soviet scientists had been regular visitors to the "sarcophagus," the shell that has been built over the damaged reactor. Many had been critical of the safety of the RBMK (graphite-moderated) units. Although no further RBMKs were being built, Pokutnyi added, those currently in operation would fulfill their terms before being dismantled.

Pokutnyi made it plain that despite international cooperation to deal with the problems arising from the nuclear accident, the tasks facing Kombinat would remain for decades. One immediate question was which ministry would be in charge of future operations. The USSR Ministry of Nuclear Energy, formed in July 1986 in the wake of Chernobyl, had been a regular target of Ukrainian activists because of its apparent indifference to republican concerns, particularly the siting of nuclear power plants. In the summer of 1989, however, its unpopular minister, Nikolai Lukonin, was relieved of his duties, and the ministry was merged with the Ministry of Medium Machine Building, which is involved with the development of atomic weapons in the USSR.

In response to the question whether it would not be preferable simply to shut down the Chernobyl plant in view of the strong opposition from ecological and populist groups, Pokutnyi stated that there appear to be two possible scenarios. Either the "green" movement will force a complete shutdown, or the nuclear industry will be forced to take every conceivable precaution to make the industry safer. On the other hand, Pokutnyi considered that the anti-nuclear lobby has been unreasonable in its attitude: emotions were playing a larger role than wisdom. Among those guilty of such an attitude, he singled out Ales Adamovich, the Belorussian writer: "He has never been here, and yet he writes constantly about these problems!"

Pokutnyi commented on the decisions to evacuate villages in Belorussia and parts of northern Ukraine, and stated that the evacuation program for such villages would begin in 1990. He also revealed that several villages within a ten-kilometer zone around the sarcophagus were used as "graveyards" for the two million cubic meters of irradiated soil collected, and for "hundreds of thousands" of tons of steel and nonferrous metals. In particular, the villages of Povesne and Lubyanka were cited. Finally, he baulked somewhat at a question pertaining to the actions of military reservists from Estonia in 1986, who were reportedly incensed when their term of work in the zone was extended from one to two, and ultimately to six months.

In fact, stated Pokutnyi, the emergency radiation norm per worker in 1986 was set at twenty-five rems, and these levels were checked on a daily basis. Although it was true that the reservists were called up for six months, he stated, if their accumulated dose approached the maximum in, say, two to three months, then they were immediately removed from the zone. Thus the first post-accident director of the Chernobyl nuclear power plant, Erik Pozdyshev, soon accumulated the maximum twenty-five rems, after which he had to quit his job, and move to work that did not require his exposure to above-norm radiation. By 1987–88, the maximum norm per worker had been reduced to five rems, and in 1990, the pre-accident rate (which was not specified) was to be restored, he noted.

THE CHERNOBYL NUCLEAR POWER PLANT

The Director of the Chernobyl nuclear plant, Mikhail Umanets, provided a more retrospective interview.[33] He focused first on the technical improvements made to the RBMK-1000 reactors at Chernobyl. Ninety fuel assemblies had been replaced with additional absorber rods, and the fuel was being enriched with 2.4 percent of uranium-235 as opposed to 2 percent at the time of the accident. As a result of such measures, it had been possible to reduce the positive void coefficient, whereby the RBMK reactor becomes unstable at low power.[34] During the interview he declared that steps were also being taken to reduce the time needed for the reactor's shutdown at unit one from twelve to two seconds—a time that compares with the Canadian CANDU reactors. In April 1986, the shutdown time was a reported twenty seconds, whereas the power surge that resulted in the explosion occurred in four seconds. This work followed experiments at the Leningrad RBMK and the Ignalina RBMK-1500 in Lithuania. In theory, therefore, if the changes were fully implemented, then it would be possible to prevent a future catastrophe similar in type to that of 1986.

In addition, reported Umanets, new safety systems were being installed, which included improved monitoring of the individual units under the supervision of the senior engineer. All the operating instructions had been revised and republished, and the personnel had been retrained or replaced. Workers from Chernobyl had been sent for training on the RBMK simulator at Smolensk, while a new, and more advanced simulator was being constructed for Chernobyl operatives in the town of Slavutych. A simulator for the water-pressurized type of reactor (the VVER) exists at the Novovoronezh nuclear power plant.

Umanets was then asked about the two unfinished units at the site, reactors five and six. In 1988, he acknowledged, he had been advocating continued construction of the station, and the bringing on line of reactors five and six. The latter are separated physically from the rest of the station at a distance of about 400 meters. Reactor five was 85 percent completed at the time of the disaster, and reactor six 15 percent completed. By mid-1989, however, Umanets had changed his mind. From a human point

33. Interview with Mikhail Umanets, Chernobyl nuclear power plant, June 14, 1989.
34. For an explanation of the positive void coefficient, see Victor G. Snell, "Introduction," in David R. Marples, *The Social Impact of the Chernobyl Disaster* (London: Macmillan), 1988.

of view, he declared, it was simply unfair to bring people to Chernobyl to construct these reactors. Indeed the wisdom of such remarks was corroborated by the radiation levels recorded on the day of the interview. At a distance of 300 meters from the sarcophagus, the recorded level was 1.6 millirems per hour, or about 160 times above the natural background. Closer to the damaged reactor, the level was said to be 10 millirems per hour, or 1,000 times the natural background.

Turning to the general situation at this time, Umanets stated that about 4,000 people worked at the Chernobyl plant, of whom only about 1,000 worked there at the time of the accident. The operatives have a thirty-six-hour work week. Most of the newcomers were sent to Chernobyl from other RBMK stations, such as Ignalina and Smolensk. About 30 percent of these workers were of Ukrainian origin, the remainder from various other republics. As to the future, Umanets appeared to be more confident than Pokutnyi concerning the public attitude toward nuclear power: eventually, he believed, it would be persuaded of the crucial importance of the industry to the Soviet economy.

On the other hand, he did comment that the days of the RBMK are numbered. A major problem has been the expansion of the graphite fuel through radioactivity, which entails replacing the fuel channels at fifteen-year intervals. In order to alleviate this problem, a major technical modification is required, and this would extend the operating span of the reactor to the regular thirty years. The future, he stated, lay with the VVER reactor type, moderated by helium.

PRYPYAT: THE "DEAD" CITY

Subsequently, I was taken by bus to Prypyat in order to visit the experimental hothouse operated by Kompleks, under the jurisdiction of the USSR Ministry of Atomic Energy (as it was called at that time). My host was a chief biologist, Borys Solomanyk. He explained that twenty people were working at the hothouse, while living permanently at Zelenyi Mys. The continuing goal of the hothouse is to examine the effect of radioactive isotopes produced by the Chernobyl explosion on various types of seasonal plants and trees. In addition to the familiar cesium, strontium and iodine, he stated, particular interest is being taken in ruthenium, zirconium and cerium. Plants are grown hydroponically in the hothouse and then taken to the open soil. He pointed out two beds of pine trees, in their second year of growth. In the first bed were "normal" trees, and the second contained trees grown from seeds collected from the "Red Forest" (close to the fourth reactor), which had received about 600 rems of radiation. The shoots in the second bed possessed elongated roots growing at strange angles. These were warped and at least three times the size of the nonirradiated shoots.

The city of Prypyat itself was deserted aside from the hothouse and a swimming pool, run by four people who were operating in shifts. Risovanny stated that in his view it will never again be populated. Two autoloader trucks were taking irradiated cars out of the town, and clearly all the vacated apartments had by then been emptied. There was a significant contrast between the confident tones of those working at the hothouse—one of whom denied that radiation had had any longterm effects upon either plants or livestock in the area—and the bleakness of the empty city, now overgrown with weeds. If any one factor symbolizes the current phobias among Ukrainians about the effects of Chernobyl, it is this deserted city of the seventies, now eerily silent.

There were a few other impressions of the zone worthy of note. First, there was a notable lack of precaution regarding protective clothing, both at the Chernobyl station itself and in the surrounding countryside where cleanup workers are on duty. Peasants with livestock were prevalent just south of the zone. Pokutnyi and his colleagues face a mammoth task in the years ahead. They are understandably anxious for good publicity. They appear to be underfunded and were facing the prospect of a serious rivalry from the Spetsatom association, which is said to be more geared toward emergencies. However, the prospects for international cooperation at the Chernobyl site appear to be good. An international Chernobyl center is being established in the town itself, while over the next few years, there are plans to build a reprocessing factory for the nuclear waste—an idea that alarms local environmentalists.

THE INTERNATIONAL CENTER

The establishment of a Chernobyl Center for International Research, located in the small village of Zelenyi Mys followed active research work in the zone by some eighty national institutes, assisted by several foreign specialists. In 1990, scientists maintained that the reactor and its environs represented a unique region for scientific inquiry. The draft program for the new Center indicated that a key supervisory role was to be played by the IAEA. The IAEA was to outline the general conditions under which research projects should be undertaken. Further, the IAEA was also responsible for publishing the results of the various projects and holding meetings and conferences under what is termed an "umbrella" agreement with the Soviet Union. Any institutions or groups that were anxious to participate in research projects had therefore to abide by this umbrella agreement and, it was implied, provide their own financing.[35]

The Center was to consist of a director and administrative, operational and maintenance personnel, all provided by the Soviet Union, along with a research council, whose members would be appointed by the Soviet authorities and were to include an IAEA representative. The USSR agreed to finance the initial expenditures, but it was suggested that, subsequently, members of projects should operate on a cost-sharing basis. Because of the potentially large number of scientific projects, no distinction was made between what could be termed "emergency projects," such as replacing the sarcophagus or building a reprocessing plant for nuclear waste, and more peripheral areas of scientific inquiry. It was also quite clear that, although the Center was located on Ukrainian territory (and only several months before Ukraine was to become an economically sovereign state), it was to be an exclusively all-Union concern.

The Center's draft program suggested seven possible scientific and technical projects:

1. Dealing with the sealed reactor, including an examination of its internal components and the migration of nuclides.
2. Decontamination work and related issues, such as the kind of techniques and equipment to be employed.
3. Decommissioning the damaged nuclear reactor, entailing an analysis of the long-term problems arising from unit four.

35. *Background Information on the Establishment of a Chernobyl Centre for International Research,* Moscow, September 19, 1989.

4. Environmental and agricultural studies of the accident's consequences, such as the movement of radionuclides in the food chain and ground water.
5. Radiation safety, including protective equipment, radiation monitoring, and the impact of radiation on health.
6. Molecular radiobiological studies on human health problems, especially radiation-induced cancer.
7. Ecology and chemistry, such as sampling and measuring methods and mobile laboratories.

There appear to be at least two major difficulties with regard to reorganization of activities in the zone that were not dealt with in the draft program. First, a new and possibly massive influx of personnel into the Chernobyl zone seemed somewhat risky from a health perspective. The zone had remained sealed, and radiation levels in 1990 were still well above normal, thus making a longterm sojourn by personnel unlikely. It is not clear why the decision was taken to establish the Center inside the zone rather than in a city such as Kiev from which periodic trips into the zone could be organized. The second problem is related to overall control of the post-accident analysis. One of the main criticisms in the Soviet Union about the post-accident studies has been that information was limited to a restricted group. The establishment of the Soviet-IAEA Center appeared to continue that policy. Access to the Center, it would seem, was to be limited to selected scientific bodies, and there was little indication that the published results would be available or—perhaps more important—comprehensible to the average Soviet citizen who has been forced to live with the accident's consequences.

There has been also a certain parochialism in the Chernobyl cleanup operation. Kombinat (or Pripyat Industrial Research Association to give the association its current name) has been obliged to concentrate on the thirty-kilometer zone. Areas that lie just outside the zone were neglected from the first. When Mikhail Umanets came to Canada in the fall of 1989, he was interviewed on CBC's "Newsworld" program, and denied knowing anything about the predicament of citizens living just outside the zone. In today's world of glasnost, this appeared to be a lamentable example of someone avoiding controversial issues. It is also a reflection of the official attitude that no serious health effects were acknowledged to have accrued from Chernobyl. In 1988–89, however, one of the most devastating episodes in the history of the protracted Chernobyl story occurred in Ukraine, namely the tragedy of Narodychi, a small raion in Zhytomyr Oblast, some sixty miles to the west of the Chernobyl plant.

3 | THE TRAGEDY OF NARODYCHI

THE MOST SERIOUS AND enduring problem to have emerged from the Chernobyl crisis in Ukraine concerns the small district of Narodychi, part of Zhytomyr Oblast, to the west of the site of the nuclear plant. In 1989, Narodychi acquired international fame through the most unfortunate of reasons, namely that it was considered by local residents to be something akin to a "coverup" operation by the authorities. If this was the case—and it can at least be demonstrated that information had not been made readily available to residents—it was one that failed. Months of pressure from local inhabitants, from visitors from Kiev, and from scientists ultimately forced the health authorities to acknowledge a problem. By the end of the year, over eighty settlements were slated for evacuation in the period 1990–92. Narodychi, then, while truly a tragedy, is also an example of the way in which popular concern has gained sway over official secrecy in today's Ukraine.

We noted above that the Ukrainian biologist, Dmytro Grodzinsky, had warned that the area had been heavily affected by radioactive fallout. It was some time, nevertheless, before Soviet officials paid attention to the question, or indeed even acknowledged its existence. Like other parts of northern Zhytomyr, Narodychi is a rural, farming region. Generally, medical facilities there are very basic, in some areas primitive. In order to examine the possible fallout, potential visitors to the zone were obliged to consult the official health authorities: either the USSR or Ukrai-

49

nian Ministries of Health, or the Kiev-based Center for Radiation Medicine.

"MI-KRO-FON!" AND ITS CONSEQUENCES

In September 1988, however, a correspondent for the Novosti Press Agency, Volodymyr Kolinko and a Kiev stage director, Heorhii Shklyarevsky made a documentary film in the area without consulting the official authorities. Entitled "Mi-kro-fon!" (Microphone), the film was in fact the second to have been made about the effects of Chernobyl, but it clearly superseded the film "Porih" (Threshold) which was produced earlier. The two producers, accompanied by Yurii Shcherbak, went directly to the local population of Narodychi, to medical centers, to farms, and asked residents about the effects of the disaster. The camera switched rapidly from the interviews to shots of calves, born without eyes and other gross deformities that were attributed to radiation. Only twenty minutes in length, the moving and powerful film fought a long battle with the censor and, perhaps in consequence, Kolinko presented his findings to the public in an issue of *Moskovskie novosti* in February 1989. This article, which caused a sensation, was then repeated by the Ukrainian youth newspaper, *Molod Ukrainy*, later in this same month.[1]

In the article, Kolinko explained that on April 26, 1986, an easterly wind blew the radioactive cloud over the Narodychi region. As a result, "today" he had found areas in which the contamination of topsoil exceeded eighty curies per square kilometer (the maximum limit for safety has been cited in Soviet works as both five and fifteen curies per square kilometer). When he had taken a Geiger counter around the areas in the fall of 1988, he had not found any area in which the radiation level was below 0.2 millirems per hour, and on some farms, the level was over two millirems per hour, compared to a normal background level in the city of Kiev of 0.014 millirems per hour. On an animal husbandry farm, which formed part of a collective farm called "Petrovsky," he had encountered levels of radiation almost 150 times those of the city of Kiev.

On this farm, the number of genetic deformities among newborn livestock had risen alarmingly since the Chernobyl disaster,

1. *Moskovskie novosti*, No. 8, 1989; *Molod' Ukrainy*, February 26, 1989.

reported Kolinko. The farm possessed 350 head of cattle and eighty-seven pigs. In the five-year period before the accident, only three cases of abnormal birth among pigs had been registered, and none among calves. Yet in the first year after Chernobyl, sixty-four deformed livestock had been born: thirty-seven piglets and twenty-seven calves. In the first ten months of 1988 there had appeared, respectively, forty-one and thirty-five grossly deformed newborns. Calves had been born, stated Kolinko, without heads and ribs, without eyes; there were pigs with deformed skulls.

Turning to the impact of the fallout on humans in this region, Kolinko noted that according to Soviet radiologists, doses of radiation of 4.4 microcuries of cesium-137 or 0.4 microcuries of strontium-90 can cause significant changes in the human organism. According to figures provided by the Narodychi Raion Medical Service, from one to two microcuries of cesium had been found in 35 percent of residents of the raion (about 37,000 people lived there, so the figure represents just under 13,000 people); from three to five microcuries in more than 4 percent, and from five to ten microcuries in just under 4 percent of residents. More than half the children in the raion had cancerous swellings of the thyroid gland, according to Kolinko. Women were reportedly inquiring why they should give birth to babies if the conditions were so dangerous. The above figures indicate that almost 3,000 residents had suffered from excessive levels of cesium alone. Kolinko declared that the number of cancers of the mouth and lip in Narodychi Raion had doubled since the time of the accident.

Kolinko's article caused a sensation in Ukraine, the Soviet Union and also in the Western media, which once again began to take a serious interest in the impact of Chernobyl. In late February 1989, Mikhail Gorbachev, accompanied by his wife, Raisa, arrived in Ukraine for a visit that included a first ever trip to the Chernobyl nuclear power plant.[2] But little new resulted from the visit. At the same time, the authorities who had been so blatantly bypassed by the Kolinko film crew were preparing a response to his findings, clearly hastened by the fact that foreign journalists

2. *Molod' Ukrainy*, February 25, 1989.

were now requesting visits to Narodychi. In fact, some had already taken photographs of the deformed mutants among the livestock. So by early March 1989, a reply had been published in the press, written by scientists from the Center for Radiation Medicine that denounced Kolinko's article as "incompetent" and unfit for export outside the Soviet Union.

The tone of the reply to Kolinko was set by a preamble in the form of a letter from Richard Wilson, professor of physics at Harvard University, who has visited Chernobyl on at least two occasions, once with a television crew from the PBS network. Professor Wilson, with regard to the Kolinko article, asked: "Is this [account] simply a case of the Soviet press's being influenced by the American tendency toward sensationalism?" The scientists then answered this almost rhetorical question with a resounding "Yes!," thereby using the foreign specialist to support their comments.

Doctor of Physics and Mathematics, I. Likhtarov, emerged as the main counter to Kolinko, and began what was to be the first of many increasingly bitter debates between the two, watched by a wary and anxious Ukrainian public. Likhtarov declared that Kolinko's article would make the hair of nonexperts stand on end. He, however, could only shrug his shoulders with amused irony upon reading the account in *Moskovskie novosti*. He immediately discounted theories that radioactive dust might have been carried great distances by the wind, arguing that a tornado or sandstorm would have been required for this to happen.[3]

Likhtarov's colleague, A. Prisyazhnyuk, who is a doctor of medicine and head of the Center's epidemiological laboratory, attempted to disprove the claim that the incidence of cancer of the mouth and lip had doubled in the Narodychi area since the accident. Because of the increased migration of the population after Chernobyl, he stated, the number of patients in Narodychi dropped dramatically to forty-nine—fewer than in 1985. In 1987, he continued, ninety-four patients were registered (presumably because some people had returned to their homes), but in 1988, the incidence of oncological diseases had fallen to seventy-four. Further, he maintained, cancers of the mouth were to be attributed to gum diseases resulting from inadequate dental work.

3. *Pravda Ukrainy*, March 1, 1989.

B.G. Bebeshko, doctor of medicine and director of the Center's Institute of Clinical Radiology,[4] maintained that more anomalies were being found among children because the number of pediatricians dispatched into the raion had risen sharply. Earlier, he declared, one pediatrician had dealt with them all, but now a large group of expert specialists were examining them. He also stated that the birth and death rates of children in the raion had remained stable in the period 1985–88. He denied Kolinko's statement that thyroid cancers had developed in almost half the children of the raion, noting that 18 percent of the children had insignificant swellings of the thyroid gland that were not life threatening. Likhtarov added that neither specialists in radiology nor the republican health authorities had recommended that women in the raion should not have children, though local doctors may have done so.

Two agricultural radiologists then commented on the alleged freak births among livestock in the raion. Academician R.A. Aleksakhin stated that it is well known that Narodychi suffers from a shortage of basic microelements in its soil, such as cobalt, iodine and copper, which can give rise to anomalies in livestock. Other possible factors, he maintained, could have been a greater application of mineral goods and pesticides, and faulty breeding techniques. He asked how Kolinko could make such a sweeping conclusion from considering only one possible cause of the mutations: radiation.

Professor N.A. Loshchilov noted that another team of specialists had followed Kolinko to Narodychi, but had failed to discern a similar number of deformities. Although he did not deny the appearance of such freaks, he claimed that, surprisingly, the section in which they were being born was one in which the radiation situation was relatively favorable. With the exception of one field on the collective farm, he stated, not one was found on which the radiation levels exceeded the permissible norm. Neighboring collective farms called "Chervone Polissya" and "Lenin" possessed higher levels of contamination of feed crops, but had not produced any deformities among livestock. He concluded that the defects among livestock on the Petrovsky collective farm

---

4. The various institutes of the Center for Radiation Medicine and their functions are described later in this chapter.

were not linked to radiation. His overall conclusion was that "close pedigree interbreeding" techniques were a more likely cause of abnormalities.

Only one of the signatories to the assault on Kolinko conceded that the situation in Narodychi might pose a danger to the population there (as later became only too evident). Doctor of biological sciences, B.S. Prister, acknowledged that there were patches of farmland in the area in which the radiation levels were around forty curies per square kilometer. This information had not been kept secret, he declared, but it did not seem to have reached the ears of the collective farm chairman. How was it possible to guarantee noncontaminated supplies if the chairmen themselves were ignorant of these patches or hotspots of radiation, he asked. Likhtarov stated that eight people out of 7,000 examined had received more than five microcuries, but that as a rule these were elderly people who had always taken milk from these cows and refused to listen to the recommendations of specialists.

Likhtarov concluded that the article in *Moskovskie novosti* was "unfit for export abroad" (most editions of the newspaper are published abroad, including the English-language *Moscow News*) and that an apology should be provided. He also attacked "Mi-kro-fon!" because of its failure to interview people who knew more about the "real situation." Likewise, Bebeshko's view was that the article and film had reduced the population's trust in radiation specialists. In closing, Likhtarov asked: "Can one achieve moral goals by immoral means," a question as rhetorical as that of Professor Wilson at the beginning of the article. However, the editorial board of *Pravda Ukrainy* stated underneath the published response, the chief problems were to be ascribed less to Kolinko and more to the absence of glasnost at the USSR Ministry of Health, which had maintained a position of great secrecy as far as information about the effects of Chernobyl was concerned.

This attempt by prominent scientists of the Center for Radiation Medicine to "set the record straight" about the Narodychi situation was soon undermined by other accounts from the area, emanating first from the region itself, and second from the information section of the Ukrainian government. While these accounts did not delve deeply into the specific state of affairs in this region, they presented a general picture that tended to support the conclusions of Kolinko. Although it might be suggested therefore that there was some attempt to mislead the Ukrainian public as to the true state of affairs by Center specialists, the generally placid picture they painted of problems unrelated to Chernobyl, and a generally healthy population clearly found some support among Western specialists acquainted with the Chernobyl situation.[5]

LIFE IN NARODYCHI, SPRING 1989

In April 1989, further evidence of a growing calamity in Narodychi surfaced in the shape of a two-part series in the Ukrainian youth newspaper by Viktor Kosarchuk and Ivan Petrenko. The two authors explained that the region had received two visits in February 1989 from members of the Institute for Nuclear Research with the Ukrainian Academy of Sciences. The visitors reported their findings to a meeting of the institute on February 22, which concluded that in the villages Klishchi, Khrystynivka and Nozdryshchi, there existed substantial areas that were unfit for human habitation. The authors laid the blame for the lack of attention to the dangerous contamination partly on the shoulders of Likhtarov. They indicated, through questions asked of a senior scientific worker at the Institute for Nuclear Research, Evgenii Korbetsky, that the information about radiation levels provided by Likhtarov was erroneous.[6]

In Nozdryshchi, the authors discovered, one woman with three children had been tending a garden which was bordered on two sides with barbed wire demarcating the "zone of alienation," the most seriously contaminated zone. In another village, they found residents cultivating land only two to three kilometers from the wire. In Khrystynivka, which had been considered a clean area hitherto, a Geiger counter revealed that the radiation level at one end of a street was tens of times higher than the

5. In addition to the comments of Richard Wilson cited above, Dr. Robert Gale, who has been very much involved in the medical aspects of Chernobyl, also expressed his strong reservations concerning radiation-related illnesses in parts of Ukraine, during an interview and discussion with this author on the third anniversary of Chernobyl. CBC Radio, "Sunday Morning," April 30, 1989.
6. *Molod' Ukrainy*, April 19 and 20, 1989.

other, yet people were ignorant of this situation. The worst situation of all appeared to be in the local forests, where radioactive hotspots had developed in fallen leaves.[7]

Following the account about Narodychi in *Moskovskie novosti*, the two journalists related, a Japanese reporter had arrived at the settlement. He was informed by the chief veterinary surgeon of the raion that approximately thirty mutant livestock had been born in 1988, in an area in which no deformities had occurred in previous years. On several collective farms in the area, he learned, it was forbidden to drink milk, or consume eggs and meat. A local postal worker apparently had pleaded with Kosarchuk and Petrenko to make sure that the children were removed from the zone. The authors provided a harrowing description of life on local farms, where the cattle were not tested for radiation, yet were being milked, and on which children were taken by their mothers into fields in which radiation levels were alarmingly high.

The newspaper of the Narodychi Raion Party Committee provided more detailed information about the problems being faced in the raion at this time. An article written by M. Atamanchuk described a packed and often raucous meeting, convoked on March 30 at the House of Culture in Narodychi to discuss the latest information about living conditions. People reportedly came from all over the raion to attend the meeting. In fact, the hall was so full that many had to listen to the speeches and questions through loudspeakers placed in the corridors, a sign of the widespread concern. The meeting was addressed by oblast and raion party and government officials, by members of the Ukrainian Hydrometeorological Committee, and by Likhtarov, the controversial spokesperson for the Center for Radiation Medicine.[8]

Various questions and comments were apparently shouted from the floor. The chairman of the raion's civil defense committee, I.P. Makarenko, testified that on the morning after the tragedy, on April 27, 1986, radiation levels in the town of Narodychi itself had reached as much as three rems per hour,

---

7. For an account of the situation in forests in northern Ukraine as a result of radiation fallout from Chernobyl, see *Lesnaya promyshlennost'*, April 8, 1989.
8. *Zhovtnevi zori*, April 4, 1989.

which would have been three times higher than the reported average for the evacuated zone. Likhtarov claimed that this figure could not possibly be accurate, whereupon Makarenko declared that not only had this figure and others not been corrected over time, but they had also been delivered to the civil defense headquarters for the oblast, in the city of Zhytomyr, where they remained today for anyone who wished to view them. Such a high radiation level clearly would have merited an immediate evacuation of the population, though it was to be several more weeks before the local party authorities were questioned over their lack of action. Local residents at the meeting were reportedly incensed that it had taken three years for any action to be taken in Narodychi even though the gravity of the situation was known to the authorities from the outset.

Citing data from the Center for Radiation Medicine itself, one participant noted that in the village of Stare Sharne (just north of Narodychi), the level of cesium contamination of the soil exceeded fifty curies per square kilometer. "How can children live in such conditions?" asked the man. "They cannot," replied Likhtarov. Another speaker asked why there had been so many contradictory statements in the press about the radiation situation. The scientist laid the blame on appraisals and prognoses by incompetent persons. "Well then," shouted another onlooker, "Can one live in Narodychi Raion or not?" Likhtarov responded in a more conciliatory manner that in twelve villages in the raion, the radiation background levels were high enough to present a hazard to the population, particularly children.

This answer evidently did not satisfy the large audience. Many described individual situations that they felt merited more urgent attention. A collective farm worker from Khrystynivka, in the eastern part of the raion, reported that in the backyard near his house, radiation levels were changing regularly by entire millirems. "But no one says anything about this. What is the point of these stupid conversations that have been continuing for nearly three years?" M.T. Kobernyk, an assistant medical officer in the raion, put it simply: "The point is that it is not safe to live in the raion. So why take the risk?" Other participants noted that the sick were not being hospitalized, that there was a "bureaucratic" attitude toward the medical observation of the population, that medical test results were not being provided, and that both the oblast (Zhytomyr) and republican health authorities had adopted a "passive" attitude toward the predicament.

The Deputy Chairman of the Narodychi Oblast Executive Committee (government), H.O. Hotovchyts, informed the audience that "at the request and insistence" of the raion and oblast authorities, a decision had been taken in Moscow to evacuate those villages now believed to be unsafe for habitation. The criterion for evacuation was that no person should be permitted to accrue a lifetime dose of more than thirty-five rems of radiation (the controversy over this issue is dealt with below). The future of the town of Narodychi itself was not examined at this time, and Hotovchyts did not feel that the situation warranted the removal of its 6,000 population.

The twelve villages in question all were suffering from cesium contamination of the soil of more than fifteen curies per square kilometer, in some cases levels were up to one hundred curies. Located in the central, eastern and southeastern parts of the raion, they were as follows: Khryplya, Khrystynivka, Mali Klishchi, Velyki Klishchi, Mali Minky, Nozdryshche, Peremohy, Poliske, Rudnya-Ososhnya, Shyshelivka, Stare Sharne, and Zvizdal.[9] In addition to the above, two villages were designated for evacuation in neighboring Kiev Oblast: Yasen and Shevchenko, both of which are located in Poliske Raion, directly to the east of Narodychi. When added to totals from the Belorussian republic, this signified that in the spring of 1989, at least thirty-five villages were now to be emptied as a result of Chernobyl-caused radiation.

THE CENTER FOR RADIATION MEDICINE

It would appear therefore that Dr. Likhtarov and his colleagues either did not know the real situation in Narodychi, or that they were unwilling to divulge this information. Moreover, the Center for Radiation Medicine was said to be too secretive. It had responsibility for monitoring the medical fallout of Chernobyl, but had declined cooperation with western authorities and individual doctors. Its director was Anatolii Romanenko, the unpopular

9. *Ibid.* The precise location of each village is cited in *Istoriia mist i sil Ukrains'koi RSR: Zhytomyrs'ka Oblast* (Kiev, 1973), pp. 438–59.

Ukrainian health minister. The Narodychi dilemma therefore seemed to owe much to the personnel of this Center, particularly Likhtarov and Bebeshko. In the summer of 1989 it was possible, however, to hold a detailed interview with scientists at this Center, and to raise some of the most controversial questions, such as the revelations about Narodychi.[10] It should be noted at the outset that while the comments may at times have been controversial, the personnel present answered all questions willingly and at some length.

The interviewees were three scientists from the Institute of Clinical Radiology: Oles A. Pyatak, Honored Scientist of Ukraine, and a Deputy Director of the Center; I.P. Los, head of the laboratory of radioecology; and V.V. Chumak, head of the immunological laboratory. Before the interview began in earnest, Pyatak provided an informative account of the division of responsibilities between the three institutes that make up the Center, which was founded on October 1, 1986 to conduct research on the effects of the Chernobyl disaster and to find methods of treating the unfortunate victims.

The Institute of Epidemiology and Prophylaxis of Radiation Exposure consists of three departments. The Department of Population Research is concerned with demography. A Registry Department analyzes data on the state of health of all the people who were exposed to radiation (in theory: the practice, according to some critics, is somewhat different as will be described below), including evacuees, those who still live close to the nuclear plant, and those involved in cleanup operations. A Department of Dosimetry and Radiation Hygiene carries out research on radiation background and tries to ascertain the levels of radiation to which people were exposed, as a guideline to future treatment.

The Institute of Clinical Radiology files data about the state of health of those people who were subjected to very high levels of radiation. Patients with visible radiation burns were much in evidence in June 1989. At the clinic and at a nearby children's unit, about 300 persons were being closely monitored. The Institute has facilities for 500 people, although at the time of the interview, there were 209 patients with diseases connected directly to

10. Interview at the Institute of Clinical Radiology, Center for Radiation Medicine, Academy of Medical Sciences of the USSR, June 15, 1989.

radiation exposure. They worked at the fourth Chernobyl unit at the time of the explosion or were firemen who came to the aid of the first victims.[11] The remainder were people with diseases unrelated to radiation but who lived in the area of control, suffering from diseases of the heart, lung and kidneys.

The Institute of Experimental Radiology conducts research on the mechanics of radiation exposure upon the individual, especially the effects of low-level radiation. At the Institute, scientists are also elaborating methods of treatment for those affected by radiation, and of special importance is research on the effects of radioactive cesium: how it penetrates the human body, how this penetration can be prevented, and how it can be removed from the human organism.

In response to a question about the routines of treatment, Pyatak stated that it depends on the level of radiation to which the patient has been exposed. The 209 most seriously affected patients must stay in hospital two times a year for up to one month. The others must attend clinics in their area of residence at least once annually. If something unusual is discovered—and it was made apparent that such has not occurred thus far—the patient is to be sent to the Center in Kiev for further examination. To date, the structure of diseases among the monitored population was said to be "normal," that is, the type of disease prevalent today is similar to that before April 1986 among all groups—newborns, children, pregnant women and adults—in the view of the Institute of Clinical Radiology's specialists.

Pyatak repeated what have become familiar figures, namely that 238 people suffered from acute radiation diseases after Chernobyl, of which twenty-eight died in 1986. Of the 238, 85 percent were said to have returned to work, in jobs in which there is no exposure to high levels of radiation. The delayed effects, he continued, were likely to be tumors, blood circulation diseases and genetic changes. Judging from the effects on the victims of Hiroshima and Nagasaki, however, such changes were to be anticipated during a time period of ten to fifteen years after the period of exposure.

11. While at the Institute, I was taken to see two of the first victims of the disaster: an operator called Symonenko and the fireman, V. Pryshchepa. Neither was willing to say much, perhaps because of the full complement of staff in attendance. Pryshchepa did remark, however, that he had been unable to work since the accident and "feels sick."

To a question about the availability of the results of such monitoring in the future, Pyatak first cited a new book, published with contributions from virtually all participants at an international conference on the medical effects of the Chernobyl disaster, held in Kiev in May 1988. The book had been available in Kiev since January 1989.[12] Soviet scientists affiliated with the Center had also delivered reports at various international conferences, he added, and had cooperated in their research with their US counterparts. In September 1989, there would be a high-level visitation from the US Academy of Sciences, while two international events had already been held at the Center under the auspices of the World Health Organization. He concluded that the Center was being completely open about the medical consequences of Chernobyl.

I.P. Los then responded to some specific questions about the situation in Narodychi, the location in which this alleged openness has been called to account. He was referred to a videofilm made in Narodychi in May 1989 called "Zapredel" (Beyond the Limits), and whether it provided firm evidence that people and livestock in the raion were suffering from the effects of high radiation. He replied that this was difficult to answer. However, "Zapredel" was already the third film on this topic (after "Porih" and "Mi-kro-fon!"). Those who were making such films, in his view, had been trying to exploit popular anxieties. While it was true that children were suffering from problems of the thyroid, such manifestations were hardly unusual, because the large swellings were considered a "normal variant" long experienced in certain areas of northern Ukraine and southern Belorussia. They were a consequence, he maintained, of a natural deficiency of iodine. If today's problems in Narodychi were compared with those of thirty or forty years ago, he continued, it would be discovered that some 30–40 percent of children at that time also had illnesses related to the thyroid gland. Then, however, the Chernobyl nuclear plant did not exist, and thus the illnesses today were a result of "natural conditions in that area."

Dr. Los was then asked about the recent assertion from Narodychi that the radiation level on the day of the Chernobyl

---

12. A.E. Romanenko, et al., *Meditsinskie aspekty avarii na Chernobyls'koi AES* (Kiev: Zdorovya, 1988). 232pp.

disaster reached three rems per hour (as noted above). He acknowledged this information, which he said was first announced by the raion's authorities. Yet, he noted, experts recognize the clear relationship between the radiation background and the effects of this radiation. If it were true that the radiation level in late April 1986 had reached three rems per hour, then the contamination of the ground with radioactive cesium today should not be the officially reported average of fourteen curies per square kilometer,[13] but 140 curies. In his view, more radiation must have remained on the ground.

Thus the current ground level precluded the alleged figures. One must consult the experts on these questions, he declared. The Narodychi authorities were well-meaning people who were trying to ease the fears of the population, but really they did not comprehend fully the situation. From Narodychi, he maintained, the Center had received no data for five days after the accident. But from data compiled on the fifth day, one could calculate the radiation background. Also, observing data compiled at Chernobyl, Kiev and Prypyat, it was inconceivable that the radiation level at Narodychi could have been three rems per hour. He admitted that radioactive "hotspots" may have been a possible cause of increased readings—such as a radioactive piece that may have dropped from the wheel of a truck—but ultimately rejected the notion that the level could have been so high, because radiation measurements must be taken at a height of one meter, not at ground level.

The three interviewees were asked why, if the situation in Narodychi is unrelated to radiation, twelve villages had been designated for an evacuation of their population. Los explained at some length that radiation is simply one factor affecting life on earth, along with physical and chemical factors. With these factors, there are three different levels of existence: normal; above-normal but harmless; and dangerous. In natural conditions, the differences between the first and second category can fluctuate by several times without having an appreciable effect on those living

13. As will be seen below, Dr. Los's figure here is higher than the officially acknowledged average given later for Narodychi Raion. While proving one point, therefore, he threw into doubt figures used to support another: namely, that radiation levels in the soil in Narodychi were not alarmingly high in 1989.

in the area. People can adjust to such new situations. Thus in regions of Argentina and India, the natural radiation background is ten times higher than the world average, but this does not bring about great changes in human existence.

Pyatak then responded to the question, commenting on the importance of psychological stress in areas such as Narodychi. Throughout the world, he said, despite minor fluctuations, there are certain standards for radiation. In 1990, the Soviet Union, which raised its maximum norms for radiation exposure after the Chernobyl accident, will lower them again. When the old standard is reintroduced, several villages would have radiation levels above this new norm. Taking into account the fears of the population, some villages would be emptied. In reality, there would be a harmless "discrepancy" between the new official limits and the actual background, which "may harm people psychologically." Radiophobia was having a significant effect on the lives of these people.

Largely as a result of the writings of media people, he concluded, a psychological tension existed among the population of Narodychi. As doctors, he and his colleagues could not separate medical and psychological illnesses, especially because the latter can quickly become transformed into the former. A conference on this subject was being planned in cooperation with the World Health Organization. But as for real illnesses, then according to current data, the appearance of future tumors and diseases would be so low that it would be impossible to determine their causes when one takes into account the natural, spontaneous development of those diseases.

In analyzing the content of this interview, it should be borne in mind that the views of the specialists at the Center—located on the northeastern periphery of the city—had become increasingly unpopular in Kiev. At the time of the interview, the two regular spokespersons for the Center, Likhtarov and Bebeshko, were attending a conference of the Academy of Medical Sciences of the USSR in Moscow. However, the comments of the three interviewees varied little from the current line. On the following day at an interview with the newspaper *Literaturna Ukraina*, I was advised not to believe the information provided by the Center. On the previous day, reporters from this newspaper had visited Narodychi and found levels of radiation of more than 200 times the natural background.

It had become clear that there was already no middle ground between two obdurate sides on the Narodychi dispute. It was also apparent that in its anxiety to assuage fears, the Center for Radiation Medicine was trying desperately to find other causes of problems that appeared to be related to radiation. One wondered why if some 30 percent of children have suffered from thyroid illnesses in northern Ukraine since the 1940s, steps have not been taken thus far to remedy the problem. More important, critics such as Yurii Shcherbak had taken photographs of the recorded radiation readings from Narodychi of April 27, 1986, which today are housed at the oblast civil defense headquarters in Zhytomyr (a detailed account of Shcherbak's views on this question and others appears in Chapter Five). For the Center's scientists to declare that they had readings only from the fifth day was simply to deny existing evidence. Were they ignoring the readings that had been taken?

The impression therefore was that to some extent the specialists from the Center, partly through insensitivity to popular fears, had lost control of the situation. Moreover, those journalists who were being blamed for arousing unwarranted fears among the population, had in fact begun their quest at the behest of those same people who no longer had faith in official statements. Volodymyr Kolinko and Andrei Pralnikov of *Moscow News,* without doubt, are journalists and not experts on radiation. They could not comment definitively on the nature of the beast. But the larger problem and the one that led directly to what has been termed here the "tragedy" of Narodychi was the restriction of official discussion to small, specialized groups, whether the international Atomic Energy Agency (IAEA) or the World Health Organization.

Thus in the major book published to date, and cited above, there is not a single contribution either by a critic of nuclear power or from any of the many observers who have questioned the official interpretation of Chernobyl. Even in the pleasant gardens of the Institute of Clinical Radiology, trees planted in memory of Chernobyl victims all bear the names of the familiar ring of officials who have monitored Chernobyl at home and abroad: Hans Blix, the Director General of the IAEA, Leonid Ilyin, Yurii

Izrael, and I. Likhtarov. All are advocates of the "lesser impact" interpretation of the disaster.

THE DEBATE OVER RADIATION EXPOSURE

This emotional debate in Ukraine on the health effects of Chernobyl had yet another aspect, namely the establishment of a new "normative" for radiation exposure in the Soviet Union, which was set at thirty-five rems over a person's lifetime to be adopted from January 1990. This figure was evidently decided upon in 1988 by the Soviet health authorities, on the advice of Leonid Ilyin, Vice-President of the Soviet Academy of Medical Sciences. Bebeshko has commented that it was reached only after due consultation with Western experts in radiation safety. In addition, he provided a description of the current radiation situation to support the thesis that an annual dose of 0.5 rems would not have any effect upon the health of those subjected to fallout.

Bebeshko was participating in an almost weekly instalment of arguments and counter-arguments on the subject of radiation exposure. He began an interview with *Pravda Ukrainy* by asserting that the radiation situation had improved considerably since the time of the accident, partly because of a drop in the capacity of short-lived radionuclides, but also because of a surprising reduction in the transfer of cesium-137 from the soil into plants.[14] As a result, the "real doses" of irradiation of the "vast majority" of the population have been shown to be substantially lower than the established norm: by three times in the zone of sanitary observation; and by ten times in the zone of acute control. In the meantime, the Ministry of Health introduced officially the new radiation limits (which had been put into practice several months earlier). The figure of thirty-five rems, emphasized Bebeshko, did not divide safe from unsafe levels of radiation, and one could exceed it by two or three times without having a negative effect upon one's health.

This total could be used, comfortably, to determine which areas were safe for habitation. Thus with a lifetime dose of less than thirty-five rems each, the population could live on a territory without restrictions. If this limit was exceeded, then decon-

14. *Pravda Ukrainy*, August 1, 1989.

tamination measures would have to be implemented to reduce external and internal radiation to an acceptable level. If this was impossible to attain, then the population would have to be evacuated. In the case of those people who had received heavy doses of radiation in the immediate post-accident days, then steps should be taken to reduce their intake in the years ahead, and to accelerate the removal of radionuclides from the organism.

Bebeshko noted that in May 1989, Soviet specialists had presented their conclusions on radiation safety to the twenty-eighth session of the Scientific Commission of the United Nations on the Operations of Nuclear Radiation, and had received general approval. Foreign specialists evidently considered that a lifetime radiation dose of even seventy-five to 100 rems would be within the bounds of safety. Similarly, representatives of the World Health Organization from Argentina, France and Canada who had visited the Soviet Union had also approved the lifetime figure of thirty-five rems. These experts reportedly enunciated the view, with which Bebeshko concurred, that the population generally and scientists who were not "experts" in radiation protection did not understand fully what was meant by a normative dose. According to Bebeshko, it was often believed that dose limits for those who worked in the nuclear power industry were the same as those that were applied in the unprecedented situation that followed the Chernobyl accident.

He also maintained that scientists untutored in radiation safety were tending to attribute all biological and medical deviations in the region of fallout to the influence of radiation. Such unfounded statements, he declared, had added greatly to psychological stress, which then adversely affected people's health and reduced public confidence in those who really were radiation specialists. Putting these comments more plainly, Bebeshko was in effect denying the right of those outside the jurisdiction of the Center to profess publicly their views on illnesses. Yet he was not opposed to the idea of a dialogue, provided that his "opponents" were as honest on the question as the specialists from the Center of Radiation Medicine.

Several experts were also called upon to support the radiation norm. The newspaper for the Kombinat staff in Chernobyl, *Trudovaya vakhta*, faced with a popular outcry against the new radiation norm, interviewed three scientists. Usher Margulis of the Institute of Biophysics at the Soviet Academy of Sciences, commented that the new dose figure was based on years of study of data taken from people who had worked in the nuclear industry and from observation of the victims of Hiroshima and Nagasaki. Also, there were territories in China and France where inhabitants could expect a lifetime dose of thirty to eighty rems without effect (the same argument that was applied during the interview of the author with scientists at the Center for Radiation Medicine). A recent meeting in Moscow of the National Commission on Radiation Safety, which included representatives of the IAEA, had agreed unanimously that the proposed norm was on the low side, and that no harmful effects could result from a lifetime dose of even fifty rems.[15]

Another scientist, Igor Ryabov, had headed a radioecological expedition in the Chernobyl area, which was affiliated with the Soviet Academy of Sciences. He referred to experiments undertaken on plants and animals in the Chernobyl zone, both before and after the 1986 accident. In the Yaniv region, for example, animals were being taken from outside and subjected to the elements for a period of time, after which they were examined for their enzymic composition and biochemical changes. Some somatic effects had occurred among them, but this result, he added, was to be expected with doses of 300–500 rems. Fish in the cooling pond of the nuclear plant had been artificially impregnated, however, and despite being subjected to accumulations of 700–800 rems of radiation, 80–90 percent of the spawned offspring appeared to be healthy. One should compare this outcome, he continued, with the 40 percent healthy offspring total in a chemically polluted water of Central Asia.

The scientists maintained that no anomalies had occurred among humans (clearly they had not visited Narodychi), because those in the contaminated territory had accumulated only about five rems in the first year of the accident, and significantly less than this amount afterward. Even children born to mothers who had accumulated 150–200 rems had not suffered from adverse health, contended Antonina Lyaginska, the head of a laboratory at the Institute of Biophysics. Consequently, these scientists were now elaborating a proposal to reintroduce agriculture in the

15. *Robitnycha hazeta*, July 28, 1989.

thirty-kilometer zone and the creation there of livestock fattening farms. In brief, therefore, the arguments advanced in these articles not only defended the new radiation norm, but even went so far as to suggest that thus far there had been no major health effects from radiation fallout.

On both counts, there were clearly strong arguments to the contrary, and they were not long in surfacing in agitated Ukraine. Yurii Shcherbak, chairman of Zelenyi svit, attacked the new radiation norm in two separate interviews. Many experts disagreed with the thirty-five rem normative, he noted, and as a medical doctor himself, he could not agree that such a dose was safe for health. In addition, he pointed out that the optimistic conclusions of certain experts that no changes had occurred in people's health or in that of nature gave rise to "serious doubts," especially insofar as the effects of low-level radiation were concerned.[16]

By the summer of 1989, Bebeshko's assertion that the radiation situation had improved over the past three years also appeared dubious in light of new evidence. In July, M.M. Kalenyk, the Chairman of the Main Scientific-Technical Department of the Ukrainian Ministry of the Forest Industry, was asked to report on the irradiation of forests in the republic. He stated that some 1.16 million hectares of forest had been contaminated, and also that the predicament had worsened over the years. In the western Ukrainian oblasts of Volyn and Rivne, for example, radionuclides had at first caught up in the tops and trunks of trees, but subsequently had fallen to the ground and were now found at depths of up to five centimeters. As a result, medicinal herbs, berries and mushrooms over a wide area of the two oblasts could no longer be collected without dosimetric control.[17]

Further arguments against the proposed radiation dose were advanced by Dmytro Grodzinsky. Aside from the argument that the experiments upon which the new normative is based concentrated on cesium and overlooked the more lethal strontium isotope, the Ukrainian biologist emphasized that budgetary restraints lay behind the thirty-five rem limit. Each time the upper

16. *Robitnycha hazeta,* July 26, 1989; and *Radyans'ka Ukraina,* July 30, 1989.
17. *Radyans'ka Ukraina,* July 5, 1989. See also *Pravda Ukrainy,* July 5, 1989, for a map on contamination of Ukrainian forests.

limit for exposure is raised by one rem, he stated, it entailed a saving of $1,000 (US) per person on safety and medical aid. He was supported in this conclusion by Leonid Kindzelsky, an expert in bone marrow transplants based in Kiev, who asserted that "This way, the nuclear industry would be free to spend nothing on ecology." The health and nuclear authorities therefore were being called to account for placing financial interests over those of the humans affected by radiation. Finally, these critics point out, with some justice, that the "trump card" of the Soviet health authorities has been the approval of the new normative by Western experts from the World Health Organization and the IAEA.[18]

The dispute over the new radiation norm fueled the crisis in Narodychi. To many citizens, the attitude of Bebeshko and his colleagues appeared arrogant and highhanded. Real fears were dismissed as emanating from "nonexperts," while the data used to support the new norm appeared at best to be incomplete. One question that had not been answered satisfactorily was how one was to deal with those who may have accumulated large doses of radiation in the first hours and days after the accident. Were they expected to remain in these same regions indefinitely when an annual dose of 0.5 rems annually would be much more dangerous for them than for the average citizen. By the late summer, residents of northern Zhytomyr Oblast, and of Narodychi in particular, were under a tremendous psychological strain

ANXIETIES AND COMMISSIONS

In August 1989, a new commission arrived in Narodychi. It consisted of a combination of newspaper reporters and a commission of biologists. Evidently, the key figure behind this new visit was Grodzinsky. In May, a meeting had taken place in Moscow between professional radiobiologists, including Grodzinsky, and workers of the USSR Ministry of Health. The latter accused the biologists of being too emotional about the state of affairs in Narodychi and not dealing with realities. In response, Grodzinsky proposed a trip there. The group consisted of twelve specialists from Moscow, Ukraine, Belorussia and Uzbekistan, but evidently did not not include anyone from the Ministry of

18. *News From Ukraine*, No. 50, 1989.

Health. Except for Grodzinsky, none had visited Narodychi previously, so it was hoped that there would be exhibited an open-minded approach and some original conclusions.[19]

The commission's trip to the area was preceded by the publication of two passionate letters to the newspaper *Literaturna Ukraina,* which had published an article about the situation on April 20, 1989. They ran as follows:

LETTER ONE

We, the residents of the village Zakusyly, are acquainted with the article which was published in *Literaturna Ukraina* on 20 April 1989, entitled "And what is the prognosis for tomorrow?" Everything that was written there about the situation that has occurred in our raion following the accident at the Chornobyl nuclear plant corresponds completely to the reality.

However, we wish to point out that for some incomprehensible reason, our village and those around it—Babynychi, Zherev, Bolotnytsya, V'yazivka, Lyplyanshchyna—have not been included in the zone that has been contaminated with radioactive aerosols. On the territory of these villages, except for local dosimetric operators, no one has undertaken measurements of radiation. We only wish that more researchers and competent specialists were concerned about these questions.

Residents of the village Zakusyly, Narodychi Raion, Zhytomyr Oblast [twenty-nine signatures]

LETTER TWO

In the name of all the residents of Narodychi Raion, we express sincere thanks to the editorial collective for the article entitled "The Truth About Narodychi." There was a lineup for two days to read this article in the reading room of the library. We hope that our raion newspaper will republish it. It is irritating for raion residents to read fabrications, though some specialists say such things and even publish them in newspapers.

During the three years that have passed since the

19. *Literaturna Ukraina,* June 22, 1989.

Chornobyl catastrophe, after which leading officials divided the raion's villages into "clean" and "dirty" (although milk was contaminated in all of them except on collective farms), residents have learned to distinguish truth from lies, and began to divide those concerned with the problems of the raion into "honest" and "dishonest."

Today, it is important to establish who is most to blame for the fact that in those villages that have to be evacuated remain people (including children): is it the former First Party Secretary of the raion committee, A.O. Mel'nyk, who was decorated with the Order of the Toiling Red Banner for his role in this affair, or someone from the Council of Ministers? For over three years in highly contaminated villages, the oblast and republican authorities, instead of evacuating people, have been opening up social and cultural amenities to the tune of 32 million rubles. Now twelve villages must be evacuated and all the available funds have been squandered on state expenses.

But we are most anxious about the state of health of our people, especially the children. The data in our possession indicate that their health is deteriorating, and the number of sicknesses has increased. However, they do not want to make note of this in the Ukrainian Ministry of Health and the Center for Radiation Medicine.

We have enough materials for yet another article for your newspaper. This theme could be continued. But we are simply sending them to you.

For your attention,
Residents of the village Stare Sharne, Narodychi Raion, Zhytomyr Oblast.[20]

Eduard Pershyn, a correspondent of *Literaturna Ukraina*, accompanied the Grodzinsky expedition to Narodychi, and spent a considerable time interviewing local residents, who by this time were said to be greeting every new medical commission with undisguised hostility. Instead they trusted their own local doctors who alongside them were having to endure the appalling condi-

20. *Ibid.*

tions of life on an irradiated land. "Among them," noted Pershyn, "the raion or village doctor is well known for his truthfulness." He described the general atmosphere in the area as one of "approaching thunder," and one in which the nerves of residents were being strained to breaking point, primarily because there appeared to be no legal or financial means of leaving the raion. The earlier investigation presided over by Likhtarov had concluded that 338 families should be removed from twelve villages, notably those with young children or women of child-bearing age.

One can estimate, using population data from these twelve villages from the 1970s, that the total number of people to be evacuated was less than 6,000 out of a total population of 25,700.[21] But the local residents wished to be moved en masse, by brigade or collective. By moving some residents and leaving others, the authorities created great tension among those left behind. Some 9,000 pensioners were living in the raion, but their pensions, as former agricultural workers, were "frozen," and thus whereas regular workers were to receive a 25 percent bonus to enable them to purchase clean food products, the elderly continued to live off contaminated vegetables and fruits grown on private plots. They simply could not afford to purchase clean food supplies.[22] Families with children were becoming frantic. One young woman lamented that:

> I was raised [on this land], it reared my children. And we eat everything that is produced on it because it is impossible to obtain anything else. Virtually no clean food products are delivered to us. It is not easy for rank-and-file workers to purchase food in cooperatives where the food prices are selfishly high. So we eat everything that the earth nurtures, even though it has been contaminated by radiation. I have four children and they are all sick. Commission after commission visits here, but all to no avail. My heart bursts with grief when I see the sick children. We are hostages. We have no opportunity to leave and yet it is impossible to live here.[23]

21. *Istoriya mist i sil Ukrains'koi RSR: Zhytomyrs'ka Oblast'* (1973), pp. 438–59.
22. *Radyans'ka Ukraina*, June 22, 1989.
23. *Ibid.*

Such concern was not restricted to Narodychi or even to Zhytomyr Oblast. One of the most disturbing facets of the Chernobyl saga has been that just as the authorities were finally prevailed upon to acknowledge one problem, another would emerge of equal import, so that from Chernihiv to Kiev, and from Zhytomyr to Rovno, few felt safe from the effects of radiation. In this same summer of 1989, it was perhaps not surprising therefore that the northwestern region of Kiev Oblast, which is even closer to the damaged reactor, should also have been suffering the aftereffects. One hundred and twenty-seven residents of Poliske Raion (Kiev Oblast) sent a letter to the newspaper *Molod' Ukrainy*, which expressed their fears for their children's future.

It transpired that three villages of this raion had been suddenly designated for an evacuation when it was discovered that contamination of the soil by radioactive cesium exceeded forty curies per square kilometer (Varovychi, Bober, and Volodymyrivka), but residents of other villages where radiation levels were said to be identical had remained. In fact, these residents had been informed repeatedly that the area was completely safe. The letter declared that when the press and television spoke about the "children of Chornobyl," they were referring to those who were evacuated from the towns of Prypyat and Chernobyl, but not about those in Poliske, who were living just outside the thirty-kilometer zone. So the residents demanded that the raion be investigated.[24] From Rovno Oblast also, anxious letters were appearing in the press about cesium contamination of the soil.[25]

A key problem is that there is no agreement on the amount of radioactive fallout that has encompassed the Narodychi area. Likhtarov, to be sure, has tended to play down the figures. He has commented that taking an area from the village of Rudnya-Ososhnya on the eastern tip of the raion (that is, closest to the Chernobyl plant) to the village of Narodychi in the center, the annual dose of external gamma radiation of the population was between 0.06 and 0.27 rems. By contrast, the raion newspaper, *Zhovtnevi zori*, reported that the gamma background of the atmosphere at Rudnya-Ososhnya was around 0.35 millirems per

24. *Molod' Ukrainy*, June 24, 1989.
25. See, for example, *Radyans'ka Ukraina*, June 7, 1989.

hour, and the soil level was 0.45 millirems. Thus the external irradiation measured by cesium alone was said to be at least ten times higher than the figures cited by Likhtarov.[26]

H.O. Hotovchyts, the Deputy Chairman of the Zhytomyr Oblast Executive Committee, has stated that six raions of the oblast were affected by radiation, of which Narodychi was the worst. They included 455 population points, with a population of more than 93,000 people, including almost 20,000 children. More than 18,000 people were living in 1989 in an area of 23,000 hectares that was contaminated, that is, the cesium content of the soil exceeded fifteen curies per square kilometer. In some parts of Narodychi and Luhyny Raion, however, levels exceeded 200 curies per square kilometer. Milk supplies were found to be at least eighty times above the permissible norms.[27]

Grodzinsky doubted whether figures supplied by the Center for Radiation Medicine corresponded with reality. Its calculations were all based on cesium he noted, "But why forget about strontium?" Transient particles of strontium-90 had fallen outside the borders of the thirty-kilometer zone and could have been the cause of the dangerous alpha-radiation that was plaguing Narodychi. Moreover, such particles could not generally be detected by Geiger counters. Much of the problem, according to the biologist, lay with radioactive hotspots. Radionuclides could have accumulated with the dimension of only a micron. Nevertheless, they could carry large doses of irradiation and were capable of destroying any living cell. These hotspots might have the capacity of "hundreds and thousands" of rems, but over a very short distance, this capacity could decrease to zero significance.[28]

If the above was the case, then perhaps there was some excuse for the authorities failure to take action. Such fallout was unprecedented and would have posed problems for any country or region. But the question that remained to be answered was why the authorities had evidently ignored the radiation readings in Narodychi and other areas at the outset, particularly those of the raion civil defense chief, I.P. Makarenko, who had reported the level of three rems per hour in Narodychi on the day after the

26. *Radyans'ka Ukraina*, June 22, 1989.
27. *Molod' Ukrainy*, August 23, 1989.
28. *Literaturna Ukraina*, June 22, 1989; and *Radyans'ka Ukraina*, June 22, 1989.

accident. This subject resurfaced during yet another visit to the region, this time by the Committee for Extraordinary Situations led by V. Doguzhiyev, which spent a lengthy portion of the summer and fall of 1989 in a further inspection of the territory.[29]

Makarenko, moreover, hardly stood alone in making such deductions. Virtually all the Narodychi authorities supported his conclusions. In August, the Doguzhiyev commission encountered a hostile audience that was demanding that those guilty of causing the predicament be punished. In addition to Hotovchyts, noted above, V.S. Budko, the First Party Secretary of the raion, was present at this gathering. He wondered why sixty-nine settlements were now officially in the zone of acute control as compared to eight at the end of 1986. Many believed that the decision to let the population remain in ignorance lay with V.M. Kavun, the First Party Secretary of Zhytomyr Oblast (though removed from office later in 1989). It was reportedly Kavun who had invested the rubles in social and cultural amenities, even while a wall map in the offices of the raion government gave a clear indicator of the real problems.

It is arguable whether officials such as Kavun knew as much as the Ukrainian party leadership. But debating where the blame lies for the Narodychi tragedy achieves little. Of more concern is clearly the flippant and often supercilious attitude of the health authorities, who had ample opportunity to visit affected villages and take measures. Had the predicament been publicized, then, as with the Armenian earthquake disaster of December 1988, it is conceivable that both domestic and international forces could have been used to combat the crisis. But how serious were the health problems that emerged: what had happened to those children who—as was remarked at the August meeting—"had been swallowing radioactive dust for three years"?

MALAISE AND MUTANTS

Narodychi first attracted the attention of the world because of the sensational mutations that were occurring among newly born livestock there. By the late summer of 1989, a grisly portfolio had been compiled by local residents, largely to counter what

29. The Ukrainian komsomol was also present. See *Komsomol'skaya pravda*, September 30, 1990.

was regarded as unjustified criticism of Kolinko and his film "Mi-kro-fon!" On farms that had produced between them a mere handful of mutants in the past, more than 200 were born with physiological anomalies in the period 1987–89: calves with three legs and without tails; pigs without eyes; a pathetic eight-legged foal. As noted, the authorities provided reasons why such deformities may have occurred that were unconnected with radiation. Nonetheless, it seems fair to point out that the scale of the problem was unprecedented and remains virtually inexplicable if one does not take into account the possible effects of radiation.

More seriously, the sicknesses among humans appeared to be increasing after 1988. They varied from sore throats and common colds, to serious thyroid problems and cataracts of the eye. In 1984, according to the chief doctor of Narodychi Raion, A.H. Ishchenko, twenty-four people suffered from cataracts of the eye, whereas in the first quarter of 1989, the figure was almost 200. By the end of June 1989, it had risen to 300.[30] In theory, all those who suffered health problems from Chernobyl were included on an all-Union register kept at the Center for Radiation Medicine. In practice, however, as was pointed out by the Kiev writer, Lyubov Kovalevska, the Center's register was far from complete. Thus while 206 evacuees from Chernobyl lived in Poliske and neighboring villages, only fifty-four of them were on the register. From Chernihiv Oblast, 1,121 affected residents were on file, but at least 1,838 had been affected by high radiation. Even more seriously, the vast majority of cleanup workers were not on this or any other register.[31]

Kovalevska provided a galling account of the death of one cleanup worker, a young military reservist from Moscow, called Leonid Ignatev, which is illustrative of the official secrecy that has surrounded the health effects of the disaster and produced situations such as that at Narodychi. Born in 1967, Ignatev was called up for service to the USSR army training center in Tallinn, Estonia. He was hospitalized briefly in April 1986, and visited by his mother, Olga. But throughout May, the mother lost contact with her son. On May 30, she heard that he was in a Tallinn

30. *Literaturna Ukraina,* June 22, 1989.
31. *Literaturna Ukraina,* August 10, 1989, & ff. The author is also in possession of an unpublished manuscript by Kovalevska on the health effects of the Chernobyl disaster as compiled from her independent survey.

hospital with a hacking cough, general drowsiness and a lack of appetite. Subsequently he was transferred for duty to Smolensk, and he called at his home along the way. His mother, in unpacking his suitcase, discovered there a commendation to Leonid for service at Chernobyl. His superior officers in Smolensk also described him as a "fine fellow" for his work in the decontamination campaign.

By July 1987, Leonid's health had deteriorated alarmingly. His hair had begun to fall out and he lost twenty-two teeth. He lived for another year in poor health before dying on July 27, 1988. Initially he was diagnosed as dying from radiation sickness, but once his body reached the morgue, doctors maintained that he had been poisoned by a little-known toxin. His army superiors have since denied that he was ever at Chernobyl, while the USSR Ministry of Defense even claimed that at the time Olga Ignatev alleged that her son was involved in decontamination work, no records were kept of the reservists who were on duty there. In short, the mother had no means of ascertaining the real cause of her son's death, though all the symptoms suggested radiation sickness.

It was as a result of instances such as the above, and similar complaints from former cleanup workers subsequently hospitalized in Moscow,[32] that Kovalevska decided to conduct a personal survey. She did so not as a sociologist or medical expert, but as an interested bystander, so that the study cannot be termed professional. It did, however, constitute the most objective examination to date of the situation. Altogether, she observed, 175,820 people were examined in the three Ukrainian oblasts most affected by fallout: Kiev, Zhytomyr and Chernihiv. Of this number, only 84,674 were found to be of a healthy disposition. In Kiev Oblast in particular, 40 percent of those treated were reported to be healthy.[33] The 1988 breakdown of sicknesses in these three oblasts was as follows:

32. See *Moscow News,* No. 31, July 30, 1989.
33. This does not signify that the remainder had all been affected by the disaster. Some Soviet specialists have pointed out that the number of healthy was so low because many of these local residents—rural peasants—had not had a medical inspection for many years. Moreover, a relatively high percentage were of an older age group. On the other hand, it would be equally remarkable if none of them had become ill as a direct consequence of radioactive fallout. See the following summary of the remarks of Yurii Spizhenko (then Deputy Minister, now Minister of Health of the Ukrainian SSR) and the con-

Cancerous growths—2,005
Endocrinal illnesses—3,046
Psychic disorders—3,051
Nervous illnesses—7,037
Diseases of the blood—36,065
Respiratory problems—21,048
Intestinal problems—11,863
Diseases of the bone
or muscular systems—9,152
Other sicknesses—9,183
Total—102,450

Before analyzing these figures further, let us examine by contrast the statements made about the health consequences of Chernobyl in this same period by Yurii Spizhenko. He maintained that the lifespan of the Ukrainian population would be unaffected by the accident, and that the maximum rise in oncological sicknesses would be an infinitesimal (and certainly immeasurable) 0.001 to 0.01 percent. He claimed that the number of sicknesses in the republic had not altered in the period from May 1986 to the summer of 1989. The ostensible rise of illnesses in areas such as Narodychi was entirely attributable to more expert methods of diagnosis and the advanced age of the population living there. Even if the population had received doses of 100–200 rems, he stated, their organisms were capable of adopting means of self-protection without medical aid.[34]

Spizhenko's theory was that many of the medical problems in the Chernobyl region today had been caused by statements of irresponsible and ignorant persons. While it may have been true that there was excessive secrecy about the medical effects of the accident, uncritical and emotional publications and speeches had done far more damage. In fact, they had led directly to increased stress, which itself brought on some of the illnesses. He cited statistics from the World Health Organization that for every one million people on earth, 1,600–4,000 are sick with cancer, and that Ukraine falls into the middle of this spectrum, with 3,000

---

clusions deduced from them.
34. *Nauka i suspils'tvo*, No. 9, September 1989.

cases annually. These comments, however, throw Kovalevska's figures into clearer perspective because the figure of 2,000 in the year 1988 would suggest that the cancer incidence in the fallout zone was already four times higher than the republican average.

Kovalevska also looked in some detail at illnesses of children, who are more susceptible to radioactive isotopes. Among the 52,000 listed on the official register in the Ukrainian areas (there would most likely be a higher total in Belorussia), 1,473 received doses of up to thirty rems of radioactive iodine; 1,177 received from thirty to seventy-five rems; 862 from seventy-five to 200 rems; and 574, from 200 to 500 rems.[35] Another factor to be taken into account was the present level of radiation in areas such as Narodychi. Taking a US-manufactured Geiger counter to the Narodychi region, Kovalevska monitored readings there of up to 4.5 millirems per hour, or over 450 times the natural background, and well above 1989 readings in the evacuated cities of Prypyat and Chernobyl.

To Kovalevska's figures can now be added others reported by Yurii Shcherbak to the founding congress of Zelenyi svit (Green World), which took place in Kiev, on October 28–29, 1989. Shcherbak disputed first of all the official data on the amount of fallout, commenting that the total was not fifty curies of radioactivity, as officially reported, but something in the region of one billion. Official data from the USSR Ministry of Health indicated that the average dose of irradiation of the population in northern Ukraine, Belorussia and the Bryansk Oblast of the Russian Republic was between five and ten rems, whereas 1.2 percent of the population had received from fifteen to 17.5 rems.[36]

Further, the figures were worse than they appeared because of the large number of young children affected. Among 1.5 million people, including 160,000 children under the age of seven, the dose of irradiation to the thyroid gland in 87 percent of the adults and 48 percent of the children was up to thirty rems. In 11 percent of adults and 35 percent of children, it ranged from thirty to 100 rems; while in 2 percent of adults and 17 percent of children, it lay between 100 and 200 rems. In this latter, most dangerous category, therefore, would have been included some

35. *Komsomols'koye znamya*, August 6, 1989.
36. *Literaturna Ukraina*, December 14, 1989.

27,000 young children. Shcherbak also revealed that the first cases of radiation-induced leukemias had broken out in Ukraine. It is impossible to reconcile such stark figures with the comments reported by Spizhenko above, particularly since Shcherbak's own figures came directly from the ministry with which Spizhenko was connected.

"THE BIG LIE"

Renewed attention was focused on Narodychi by the Fall 1989 session of the USSR Supreme Soviet. It had been reported at this same time that children in at least two of the twelve villages slated for evacuation had started to attend the regular school session, a fact that some critics considered an outrage.[37] On October 19, three committees of the Supreme Soviet discussed the state of health of the people in radiation-affected zones. The committees in question were the Committee for Public Health, the Committee for Ecology and the Rational Use of Natural Resources, and the Committee for Women's Affairs. It was acknowledged that for a lengthy period, medical experts had not provided an objective figure of the health effects of Chernobyl. Members of these three committees therefore held a meeting at the USSR Ministry of Health, which evidently went some way toward convincing critics like Shcherbak—here in his capacity as Chairman of the Subcommittee on Ecology—that the voice of official medicine was at last dealing in realities.[38]

Shortly before the meeting with the health authorities, however, animosity over the Zhytomyr situation had reached a new peak. Shcherbak, Ales Adamovich, Valentyn Budko, Kolinko, Alla Yaroshinskaya, a journalist and people's deputy, and others took part in a discussion at the editorial offices of the weekly newspaper, *Moskovskie novosti,* and the resultant two-page account was entitled "The Big Lie." Using as its focus the radiation picture in the Belorussian republic and in Narodychi, the participants asserted that from the outset, there had been an official coverup of the effects of Chernobyl, and that this concealment constituted a crime. For over three years, the population living in

37. *Molod' Ukrainy,* October 12, 1989.
38. *Radio Moscow,* October 19, 1989.

these areas had been kept in ignorance of their environment. The participants singled out as principal culprits, the two republican party leaders at the time of the disaster, Volodymyr Shcherbytsky in Ukraine, and Nikolai Slyunkov in Belorussia.[39]

In addition, it was alleged, the victims of the disaster had been systematically concealed also. Yaroshinskaya provided several examples of the classification of information about radiation effects, and showed how the Government Commission that was appointed to deal with the consequences of Chernobyl, led by Borys Shcherbyna, had refused journalists access to information. Adamovich and Kolinko demonstrated how radioactive food products were not only being grown locally in radiation-affected zones, but continued to be distributed to markets throughout the country. The discussants remarked that once a lie is initiated, then it is inevitably made worse, the consequence of a flawed and bureaucratic system to which is subordinated science and medicine.

In the villages themselves, residents were starting to turn to sources that they felt would publicize honestly the state of affairs. One such source was the newspaper *Molod Ukrainy*, which was inundated with letters describing the "three-year coverup" in the northern regions of Zhytomyr Oblast. Some of these letters demanded that those responsible must be punished, but a growing number were now from outside the problem area, from people who offered assistance to those in need.[40] Angry sentiments were expressed by members of the Central Committee of the Ukrainian Komsomol organization, which demanded the immediate evacuation of residents in affected raions of Zhytomyr, Kiev and Chernihiv oblasts, and at the same time sent squads of people into the Narodychi area to offer help on a daily basis.[41]

FROM NARODYCHI TO LUHYNY: SPREADING CONCERN

By October there was more disturbing news from this besieged region, with the announcement that a further seventy-three popu-

39. *Moskovskie novosti*, No. 42, October 15, 1989. See also the article by Adamovich in issue No. 41, October 8, 1989.
40. *Molod' Ukrainy*, October 10, 1989. On this same topic, see also *Molod' Ukrainy*, September 17, 1989; *Komsomols'koe znamya*, September 17, 1989; and *Molod' Ukrainy*, September 23, 1989.
41. *Molod' Ukrainy*, October 10, 1989.

lation points were to be evacuated over the next few years, rather than the twelve initially anticipated.[42] In late November, new reports came from other regions of Zhytomyr Oblast. A radiological engineer from the Luhyny Raion, called S. Vasilyuk, wrote an angry letter to a Kiev newspaper to complain that the situation in that raion had been misrepresented by V.K. Chumak, the Director of the Center for Ecological Problems of Atomic Energy, affiliated with the Ukrainian Academy of Sciences. Chumak and a team of scientists had visited the raion and established that only nine of fifty villages were suffering from cesium contamination of the soil that was over five curies per square kilometer. Vasilyuk pointed out that data from the research station in Zhytomyr revealed that more than 50 percent of the raion's arable land exceeded this figure, encompassing twenty-seven population points.[43]

There was said to be a similar discrepancy with the radionuclide content of milk. According to the Chumak inquiry, only in two villages did tests of milk indicate levels of contamination of more than 50 percent, while the average for the remainder was said to be 36.8 percent. Yet the raion's bacteriological laboratory had conducted probes which indicated that 80 percent of milk samples from the private sector were found to be over the maximum limit for radionuclide content, in some cases by more than twenty-four times. The cesium content of vegetables and potatoes was said to be below normal in the Center's report, but Vasilyuk maintained that the procurement office had refused to accept these crops from local farms because they were unfit for consumption.

Most controversial was the individual and external irradiation of Luhyny residents. Chumak provided an average figure of 0.28 rems per year, which would have fallen below the 1990 official lifetime limit of thirty-five rems over a seventy-year period. By contrast, laboratories under the jurisdiction of the Ukrainian Ministry of Health had established that residents of this raion would likely receive on average two to three rems a year, or up to six times the maximum permissible (leaving aside the accumulations of the immediate post-Chernobyl era). Vasilyuk stated

42. Radio Kiev, October 23, 1989.
43. *Radyans'ka Ukraina,* November 24, 1989.

that the real figure was considerably higher than this level, and noted that local residents had begun to complain only now rather than earlier because it had finally been made evident that measures taken to date to alleviate their predicament had been woefully inadequate.

In a commentary on this letter, L. Brovchenko, a correspondent for the newspaper, informed that when he attended a meeting in Luhyny, at which was present both Chumak and local residents, it was as though an invisible barrier lay between the two sides: "One attacked, the other defended." The correspondent "sympathized sincerely" with the population of Luhyny Raion, especially in a situation in which no definitive conclusions had been reached about the dangers. But all interpretations of the Chernobyl radioactive fallout, he concluded, must have as their common denominator, the truth. However, the difficulty has been that those investigating the radiation conditions in Zhytomyr Oblast had rarely been willing to publicize the results.

In addition to Narodychi and Luhyny raions, Korosten Raion, also in Zhytomyr Oblast, was facing a crisis situation by the end of 1989. Korosten, a town of around 75,000, was one of numerous "new settlements" that had begun to appear on the radioactive fallout "map."[44] Its problems were discussed during a visit there of K.I. Masyk, the Chairman of the Permanent Extraordinary Commission of the Ukrainian government, which was investigating the effects of Chernobyl. As in Narodychi, people were deficient in supplies of noncontaminated food, particularly fruit and vegetables. It was also reported that local villagers were obliged to wipe their hands in clay, since there was no soap available; nor were there public baths.

V.M. Benya, the First Party Secretary of Korosten Raion, stated that "Our problems cannot be put off until tomorrow." Children's health was the primary concern, but the raion did not even possess a hospital. Those who had to work permanently in Korosten, in his view, should be permitted to take sixty days of vacation, and in villages where radiation levels posed a hazard to health, then the population should be evacuated. I.M. Chyryk,

44. In 1972, the estimated population of Korosten Raion was 123,200. A rough estimate of the 1989 population, based on the average republican increase in the period 1973–89, would be around 164,000. See *Istoriya mist i sil: Zhytomyrs'ka Oblast*, p. 307.

First Party Secretary of the city's party committee, added that all the residents were bringing their concerns to the authorities, "but we are not all-powerful." The party itself was basically ignorant about the true situation, and did not dare to mislead citizens for fear of being unable "to look them in the eye" the next day.

One correspondent who reported on this meeting, V. Skoropadska, provided a personal footnote to the effect that difficulties such as the above had too often simply been dismissed as part of "radiophobia." She recalled, however, the haunting words of a woman at one of the many meetings in affected villages of Zhytomyr Oblast:

> I am still young. I want to live, and I want my children to live. But when I give my children a glass of milk, I feel treacherous, because I do not know how much cesium it contains. I have already been in hospital three times, and so have my children. Yet before the Chornobyl tragedy, we were all healthy.

As the quotation indicates, the new revelations had fueled more psychological tension in the villages of northern Ukraine. Skoropadska maintained that to blame the local population for the onset of radiophobia was offensive. Rather, the forces of the Ukrainian republic, and preferably of the entire country had to be mobilized to resolve the crisis, and expenses should not be spared. The latter, in her view, had become immaterial because one was dealing with the wellbeing of a future generation.

The informal Ukrainian ecological association Zelenyi svit had begun to take an active unofficial role in investigating the extent of the radioactive fallout. In the late summer of 1989, it commissioned members from the Institute of Nuclear Research to visit the affected raions. Evgenii Korbetsky was one of the members of the group. He revealed that when the leaders of the Center for Radiation Medicine heard about its formation and intentions, they at once made a telephone call to the institute, demanding to know who had given permission for employees of the institute "to interfere in places where they are not needed."[45] There could hardly be a more illuminating indicator of official paranoia over

---

45. *Robitnycha hazeta*, November 16, 1989.

releasing information about the situation in Narodychi and its environs.

Korbetsky provided a bitter critique of what he called the "Ilyin theory" on the effects of Chernobyl. Ilyin, the Vice-President of the USSR Academy of Medical Sciences, had predicted that Chernobyl would have only a small impact on the health of people who have lived under the path of the radioactive cloud. He has also supported the thirty-five rem radiation norm for Soviet citizens. Korbetsky stated that this latter viewpoint contained far too many loopholes, not least ignorance of how much radiation the residents of Zhytomyr Oblast actually received in 1986. Moreover, Ilyin had assumed that inhabitants had consumed only clean products since that time, which was untrue. "We consider," he ended, "that the residents of Narodychi Raion have already received ten rems."

Zelenyi svit offered several proposals. First, that any residents in contaminated regions who wished to leave should be permitted to do so if the cesium content of the soil exceeded five curies per square kilometer. Second, those families with children should be evacuated "in the current year" (it seems that the meaning here was within a one-year period, that is, by November 1990). Third, that those who remained in the Narodychi Raion, and were working in areas where the contamination of the soil was more than two curies per square kilometer (and with the addition of pensioners) should be placed on the same footing as professionals who worked regularly with ionizing radiation in terms of hours of work, payment, and vacations. This provision was partly in response to the impecuniousness of villagers who had been thereby prevented from buying clean supplies of food from cooperative and state stores.

In November 1989, at a session of the CC CPSU Politburo devoted to the problems of Chernobyl, Hryhorii Revenko, then First Party Secretary of Kiev Oblast, provided an example of the ambivalence with which the Ukrainian authorities have reacted to the new information about the effects of Chernobyl. On the one hand, he acknowledged that a serious predicament existed on one-third of the territory of Kiev Oblast, which possessed a population of about 300,000, including 59,000 children. On the other hand, he exhibited the same attitude toward "unofficial" observers as the specialists from the Center for Radiation Medicine. The "Greens" and members of the Ukrainian Popular Movement in Support of Perestroika (Rukh), he remarked, were actively involved in destabilizing the situation by spreading rumors and raising emotions and anxieties in the unfortunate areas.[46]

THE FOURTH ANNIVERSARY AND BEYOND

The replacement of Anatolii Romanenko as Ukrainian Minister of Health with Yurii Spizhenko in November 1989 resulted in a new frankness about the effects of Chernobyl in Ukraine. Spizhenko, as noted, had been somewhat reticent hitherto and had largely confined himself to supporting the official line: that there had been no discernible health effects thus far. However, once in his new position, he cooperated with the Western organized Children of Chernobyl association (a similar Soviet organization was scheduled for September 1990), and informed the Soviet authorities in Moscow that one million Ukrainians had been affected by radioactive fallout, while some 200,000 cleanup workers had required or were requiring medical attention. It was also revealed that 3.5 million hectares of agricultural land in the republic had been contaminated by radioactive cesium and 1.5 million hectares of forest.[47] While this still was hardly the full story—radiostrontium is known to have had a profound impact on Ukrainian agriculture, for example—it was, in the view of critics, a definite step in the right direction.

By the time of the fourth anniversary, in April-May 1990, certain features of the post-accident analysis had become familiar. Most notable was the gap between the conclusions of a variety of scientific commissions, and the realities of the accident among the population at large. Scientists had thus concluded that in certain cases, radiation levels had been too low to have had significant health effects, but at the same time in many villages there had already been a discernible and alarming rise in oncological sicknesses. Such a tendency is not uncommon. It resembles the sort of ivory tower image of experts who have based their con-

---

46. *Pravda*, November 11, 1989.
47. See David Marples, "One Million Ukrainians Affected by Chernobyl," *Report on the USSR*, Vol. 2, No. 13, March 30, 1990.

clusions on risk estimates rather than direct contact with the population. An example of this approach was provided by the second All-Union scientific meeting on the "results of liquidating the consequences" of the Chernobyl accident, which took place in late May 1990 in the town of Chernobyl. Some 400 scientists participated, from all parts of the Soviet Union.[48]

In fairness, some speakers adopted a realistic approach to the immense health problems. K.K. Dushutin, the Deputy Director of the Pripyat Research and Industrial Association, for example, pointed out that the health of cleanup workers was likely to be affected by premature ageing processes, a rise in nervous diseases, heart vessel illnesses and digestive problems. The longterm effects of ionizing radiation, in his view, would possibly be revealed less in physical illnesses than in the reduction of the natural lifespan. Such a conclusion is supported by Zhores Medvedev, who had noted in his book on Chernobyl that a radiation dose of 100–200 rads might shorten the life of a person by six or seven years.[49] Dushutin's comments, however, contrasted with those of N.N. Savkin, head of a laboratory at the Institute of Biophysics, who had been involved in research for the previous four years on the evacuated zone, thirty kilometers in radius around the damaged reactor.

Savkin maintained that there was a real possibility of returning evacuees to the "zone of alienation" to live a normal life. Assuming that evacuees returned to the zone in the year 1990, over a period of seventy years, they would receive only between 1.5 and 6 rems of radiation, and could live without restrictions, workers in Savkin's laboratory have determined. It appeared that this view did not meet with unanimous approval, but many of those present at the meeting concurred that shiftwork in the zone on a longterm basis was acceptable. Other speakers objected to the fact that the republics had assumed control over aspects of scientific research on the affected population, especially in the nonevacuated "zone of acute control." One declared that while the Kombinat/Pripyat association was financing such work, the independence of researchers was assured, but now this associa-

---

48. *Robitnycha hazeta*, June 2, 1990.
49. Zhores Medvedev, *The Legacy of Chernobyl* (New York: W.W. Norton), 1990.

tion could control only the monitoring of those workers directly under its jurisdiction. There was also discussion of how one could control the passage of radionuclides into the organisms of livestock to produce "clean" meat and milk.

In some respects, therefore, the meeting returned to the heady days of optimism of 1986; on how to survive in a contaminated territory, and scientists associated in particular with the Institute of Biophysics of the USSR Ministry of Health objected to the "control" of their work by the Belorussian, Russian or Ukrainian authorities. By contrast, one could observe an apparent spreading of the effects of radiation across Soviet territory, ostensibly not just because the danger of radiation increases with time, as radionuclides penetrate the food chain, but because of the failure to discern the fallout region at the outset. In the summer of 1990, there remained several regions of concern.

NORTHERN UKRAINE

On the date of the fourth Chernobyl anniversary, the newspaper *Pravda Ukrainy* published an interview with Deputy Chairman of the Ukrainian Council of Ministers K.I. Masyk, who had been the Ukrainian government official assigned specifically to deal with the effects of Chernobyl.[50] He noted that 1,614 settlements in Ukraine had been irradiated, which were inhabited by 1.44 million residents (a total that superseded the one million cited by Spizhenko), including 250,000 children. In regions such as Narodychi, Ovruch, Olevske and Luhyny, 150,000 people had manifested signs of severe irradiation from radioactive iodine. In a separate survey, cited by Green Council member Mykhailo Sydorzhevsky, more precise figures on illnesses attributed to radiation were provided by B.A. Korzhanivsky, the Chairman of the Public Group for Radiation Protection of the Population, which were divided into two sections: the first for adults and the second for children.[51]

Among adults, comparing statistics for the years 1986 (before the effects of the disaster would have had time to take a major effect) and 1989, the number of those sick with specific illnesses

---

50. *Pravda Ukrainy*, April 26, 1990.
51. *Zelenyi svit*, No. 1, April 1990, pp. 4–5.

in Narodychi Raion were as follows. For those of the stomach-intestinal tract, 851 and 1,775 respectively; lung diseases, 740 and 1,275; hypertonia, 1,556 and 3,090; chronic heart disease, 764 and 2,715; and for vegetative dystonia, 0 and 442. Among children, against listing first the figure for 1986 and second, 1989, for thyroid tumors at the secondary stage, the numbers were 22 and 433; for tumors at the third stage, 0 and 29; for sicknesses of the ears, throat, and nose, 101 and 481; eye diseases, 47 and 331; sicknesses of the digestive organs (liver, stomach, kidneys, stomach gland), 14 and 491; lung diseases, 138 and 2,079; anemia (low blood count), 17 and 510. In brief, there had been an undeniable and significant rise in sicknesses among the population in a zone known to be adversely affected by high levels of radiation, and within three years. Such figures rendered dangerously superfluous, prognoses about longterm risk coefficients.

In February 1990, V.G. Fedorov, the First Party Secretary of Zhytomyr Oblast, had provided an update about the situation there. He maintained that over the previous four years, 900,000 hectares of land in the oblast had been contaminated; this constituted 50 percent of its total territory, encompassing two towns, eight raions, and 577 villages, with a total population of 362,000. He was irate that the government and state commissions responsible for inspection of the soil had been excessively slow in carrying out their task. The soil in Korosten, for example—a city of about 72,000 people—had not been examined before 1990. When it was finally tested, thirty of the fifty probes revealed cesium levels that exceeded the permissible norm. What, Fedorov wondered, had the state and republican commissions been doing for the previous four years?[52]

By mid-April 1990, Yurii Izrael, a controversial Chairman of the USSR State Committee for Hydrometeorology and Environmental Control, acknowledged in an interview in *Pravda* that the radiation contamination of the soil in parts of Narodychi and Poliske exceeded 100 curies per square kilometer.[53] About 15,000 people were to be resettled quickly on the basis of a norm of forty curies (evidently an emergency norm), in addition to some

52. *Radyans'ka Ukraina*, February 20, 1990.
53. *Pravda*, April 17, 1990.

families who were said not to be living in such dangerous areas, but who had expressed a desire to leave. Some 2,053 families were to be evacuated from Poliske Raion (Kiev Oblast) and 936 from Narodychi by 1993, from nineteen settlements. Among these evacuees were to be families with pregnant mothers or schoolage children. They were to be housed in various parts of the republic: city dwellers were to be sent to towns and rural dwellers to the countryside. Four hundred apartments were being vacated "willingly" by Kiev citizens for new evacuees. A further 59,000 residents living in this zone were to have the opportunity to buy 5,000 new homes outside the zone over the next two years.[54]

THE QUESTION OF SLAVUTYCH

In 1986, construction had begun on a new city for nuclear plant operatives called Slavutych, about 40 miles to the northeast of the Chernobyl station. The anticipated population was 30,000 for what was described as a "21st century city," with all modern amenities and with an electric train service to the nuclear plant. The city was duly completed in foundations and in the summer of 1990 held a population of around 20,000. However, when the republican governments of Belorussia and Ukraine published maps of radiation in March 1989, it was revealed that Slavutych itself appeared to lie in a radioactive patch. During the interviews conducted at Chernobyl in the summer of 1989, a leading official of Kombinat (Pripyat) association declared emphatically that this was not the case, because the site for the new town had been selected with great care.[55] But the fourth anniversary reports negated this statement.

An interview was held with V. Zhigallo, Chairman of the city government, who noted that by next year, the population was scheduled to rise to 30,000. However, Slavutych had in fact been located, it was reported, in the most contaminated area in the region and "no one can explain" how this occurred. The original location had been approved by S. Burenkov, USSR Minister of

---

54. *Pravda Ukrainy,* April 26, 1990.
55. Pavel Pokutnyi, Chairman of the Department of Information and Foreign Relations, during an interview on June 14, 1989.

Health at that time, and by Yurii Izrael, and subsequently the feasibility of the site had been reconfirmed by Izrael and new health minister, E. Chazov. However, while scientists concurred that the town was safe, forest workers felt otherwise and began to put up notices warning the population to take precautions against radiation. In 1989, over fifty million rubles had to be expended in removing more than 200,000 cubic meters of irradiated topsoil in the town, but the surrounding forests had remained severely contaminated. The city population was living on emergency norms of thirty-five rems of radiation per person, and soil contamination of forty curies per square kilometer.[56]

Life for city residents was one of great uncertainty. It was possible to change one's residence, especially if one had been working in a zone of high contamination, but difficulties had arisen because of the unwillingness of outsiders to exchange present apartments for those in Slavutych. Each family member had been allocated fifteen rubles per month to purchase noncontaminated food; and an unspecified number of Geiger counters were said to be available. But illnesses abounded and though their origins were not always clear, there appeared to be an extremely high proportion of problems in childbirth. Of the sixteen children born in the city in the first quarter of 1990, three were stillborn and five were seriously ill. There were thirteen cases of leukemia among city children. Added to such problems was the insecurity of some 300 specialists and thousands of support staff who were to be without jobs after 1995, when the Chernobyl station was to be decommissioned.[57]

THE SITUATION IN BELORUSSIA AND RUSSIA

While the focus of this work is on Ukraine, it would be misleading to ignore the effects of Chernobyl elsewhere in the Soviet Union. The lengthy article published by Yurii Izrael in *Pravda*, which was to some extent an attempt at self-exoneration in the

56. *Rabochaya trybuna*, April 26, 1990. Hotels and other facilities for the new international center being established near Chernobyl were being built at the shift settlement of Zelenyi Mys, just outside the thirty-kilometer zone. See *Robitnycha hazeta*, May 30, 1990.
57. See *Radyans'ka Ukraina*, May 20, 1990; and *Robitnycha hazeta*, May 31, 1990.

face of adverse criticism (from Yurii Shcherbak, the Belorussian deputy Ales Adamovich, and others), did reveal that the consequences of the fallout had been more severe in Belorussia than elsewhere. A Western critic pointed out in a lecture of mid-1990 that while Belorussia received a minuscule proportion of the funds allocated for evacuations and decontamination, two-thirds of the fallout from the disaster landed on Belorussian territory.[58] Izrael's figures suggested that of the 1,670 square kilometers of territory contaminated with over forty curies per square kilometer, 1,160 lay in Belorussia, 310 in the RSFSR, and 200 in Ukraine, excluding the zone already evacuated (inclusion would have rendered Ukraine the second worst affected area). A map published alongside Izrael's article showed that the fallout zone in the Belorussian republic extended not merely north of Gomel, but almost as far north as the city of Mogilev, dispelling premature conclusions that only the republic's southern territory was affected.[59]

A particularly problematic area was the forests of the southern part of Gomel Oblast, especially around Khoiniki. In mid-1988, the Belorussian Academy of Sciences and the State Committee for Hydrometeorology of the republic declared about 142,000 hectares to be a "radioactive reserve." It was estimated that half of the radioactive fallout in this area fell into the woodland. By early 1990 there had been noted changes in leaf sizes and fauna, high levels of radioactivity in hedgehogs, birds that lived near water, and also in animals that migrated into areas well away from the zone of major contamination, such as foxes and hares. However, according to a local official, the more time that had passed since the Chernobyl accident, the less concern officials had shown for human suffering that was occurring in the Belorussian republic.[60]

One area of Mogilev, it was reported, was located 200 miles from the Chernobyl plant, but had been completely resettled in

---

58. Zina Gimpelevich-Schwartzman, "Byelorussia: The Fourth Year After Chernobyl," paper presented at the annual conference of the Canadian Association of Slavists, University of Victoria, June 1, 1990.

59. *Pravda*, April 17, 1990. Using a criterion of forty rather than fifteen curies was also misleading. In the past, the latter had been used as the basis for a contaminated region, and would naturally have rendered the affected zone—especially in Ukraine—far larger than stated by Izrael.

60. *Lesnaya promyshlennost'*, No. 12, January 27, 1990.

summer 1987. The average level of cesium-137 in the soil was 146 curies per square kilometer, and in a local pond, the instruments registered an astonishing 3,000 curies. However, there were many other areas with less spectacular but still dangerous levels of contamination that were not being resettled. Altogether, 18 percent of the agricultural land of Belorussia had been severely contaminated by the Chernobyl accident.[61] In 1990, some 3,000 families in the Mogilev region had to be resettled. In an area with a population of 2.2 million, it was claimed, 20 percent had been seriously affected by Chernobyl, and the republic was in need of eight billion rubles merely to preserve the health of the population in affected regions. The population was said to be increasingly bitter and a private list of those who were believed to have died from radiation sickness had been drawn up.[62] In mid-1990, while at least something was now known of the tragedy of Narodychi and northern Ukraine, the wider picture in Belorussia was only just emerging.

As for the Russian Republic, a total of 8,130 square kilometers possessed a cesium content of more than five curies per square kilometer in mid-1990. In a high-level report about the current situation in Russia, V. Doguzhiyev, who had headed a group of experts reexamining the effects of radiation, cited the Bryansk Oblast as the only contaminated region of Russia. Deputy V. Samarin, however, pointed out that maps now available showed that Kaluga, Orlov, Tula and Kursk regions also had to be added to the "radioactive map." He also maintained that in these areas, sicknesses caused by radioiodine to the thyroid gland had risen by up to five times among the population generally, and by eight to twelve times among children, since 1986.[63] Some thirty-one villages in the Russian republic were to be evacuated in 1990–92, but this appeared to address only the periphery of the problem.

PROGRAMS OF ACTION

The Green World ecological association (Zelenyi svit) had long advanced its ideas on how the problem of Chernobyl should be

61. *Ekho Chernobylya* (Minsk), No. 1, April 1990.
62. *Rabochaya trybuna*, April 26, 1990.
63. *Izvestiya*, April 26, 1990.

approached. In February 1990, it was able to advance some of them at a meeting of the Ukrainian Supreme Soviet on the ecological situation in Ukraine. There was in place a so-called "Complex Program" drawn up by the Ukrainian Academy of Sciences, the Center for Radiation Medicine, and the party and government organizations of the affected oblasts. But it appeared that with each passing month, the actual radiation situation appeared to be worse than anticipated. Yurii Shcherbak, the Chairman of Zelenyi svit, was not merely dissatisfied with the actions taken after the accident, he demanded that a special parliamentary commission be set up to investigate the "criminal concealment" of information from the public of the Ukrainian, Belorussian and Russian republics. The commission would study not only decisions made at the highest level, but would focus particular attention on the actions of Borys Shcherbyna, as chairman of the government commission appointed to investigate the accident; those of Yurii Izrael; and above all those of Leonid Ilyin, Vice President of the USSR Academy of Medical Sciences.[64]

Ilyin, in Shcherbak's view, was guilty of criminal offenses. He had allegedly prevented the prompt evacuation of children from Kiev in early May 1986; he had taken steps to hinder the dissemination of objective information about the situation after the explosion; and he had instituted the "socially treacherous" new radiation limit of thirty-five rems per person over a seventy-year lifespan. Shcherbak pointed out that Anatolii Romanenko, the current Director of the Center for Radiation Medicine, and his colleagues, such as Ilya Likhtarov, had been the targets of serious accusations at a number of meetings in the republic. For these reasons, he stated, he supported a proposal advanced by Yurii Spizhenko, the new Ukrainian Minister of Health, that the Center for Radiation Medicine be removed from the jurisdiction of the USSR Academy of Medical Sciences (that is, Ilyin) and be placed under republican jurisdiction.[65]

Shcherbak also called for the release of information on the true scale of Chernobyl; immediate action regarding 800 sites

---

64. *Radyans'ka Ukraina*, February 20, 1990.
65. It should be emphasized that Shcherbak was one of many speakers who angrily criticized Professor Il'yin, who had become something of a pariah for his frequently flippant and condescending statements on the "minimal" consequences of the disaster. See *Radyans'ka Ukraina*, February 20, 21 and 22, 1990.

used for the temporary burial of radioactive waste; the creation of a Ukrainian national commission for protection against radiation with the exclusive right to establish the levels of radiation that might be injurious to health; the solicitation of international assistance, including the new scientific center; the mass production of Geiger counters; and the issuance of a decree halting construction of new nuclear reactors on Ukrainian territory. Despite such a fiery speech, the actual decree adopted—and it should be recalled that this Ukrainian Supreme Soviet predated the 1990 elections—was vague and even confusing because it did not treat Chernobyl separately from other ecological issues. It did, however, demand that the Center for Radiation Medicine and other institutions be transferred to the jurisdiction of the Ukrainian Ministry of Health.[66]

A new "united" Program to eliminate the consequences of Chernobyl was published in late April 1990 by the USSR Supreme Soviet, which in its preamble noted that existing measures had been "inadequate" and that there was a need to develop a program based on longterm cooperation between the central government and those of Belorussia, Ukraine and Russia. Tens of thousands of people, it was stated, were then living in contaminated regions, and they required both clean food and medical attention. Therefore the Supreme Soviet had issued a six-point program to "liquidate the consequences" of the accident in 1991–92, and to resettle families, especially those with pregnant women and young children, using funds from the all-Union budget.[67] The plan came into force in late 1990, and a "Law on the Chernobyl Catastrophe" was to be presented to the USSR Supreme Soviet in the last quarter of 1990. There were five basic problems to be considered:

1. Improvement of medical services in affected regions, especially for very young and schoolage children.
2. The development of "sorption" methods of preventive medicine and dietary methods of removing radionuclides from human organisms.
3. The development of new approaches to reduce radioactivity in

66. *Radyans'ka Ukraina*, March 1, 1990.
67. *Izvestiya*, April 27, 1990.

areas and to prevent the spread of radionuclides to new regions.

4. The development of a unified governmental system of benefits and compensation for those affected by and living in the zones of the Chernobyl accident, as well as those involved in the decontamination process. A state register was to be compiled of these people.

5. The creation of a unified system of radioecological information which would report on radiation levels to the general public. All secrecy on this question was to be abolished.

Other major points in the Program related to the need to take the nuclear plant out of commission and to establish a "Children of Chernobyl" Program by 1990. In 1991, maps were to be drawn up that would indicate the "exact" extent of the affected regions. April 26, in Belorussia and Ukraine, was designated as a national day of mourning: "the day of the Chernobyl (Charnobyl, Chornobyl) tragedy."[68]

CONCLUSION

The Narodychi story, and that of northern Zhytomyr Oblast in particular, still has to run its course. It has seen the unravelling of a web of deceit woven by official sources; one that might have succeeded at an earlier period of Soviet history, and perhaps even in the Gorbachev period but for the Chernobyl disaster itself. It is arguable that had the authorities been forthcoming from the first about the true health consequences of the accident, then much of what has happened since may have been avoided. The development of a variety of informal groups owed much to what was perceived as official secrecy. Journalists who had been roundly berated for daring to step into the "holy preserves" of science soon reacted in the logical way by employing the help of other scientists who were able to piece together an alternative picture—the roles of Grodzinsky and Korbetsky have been illustrated in the above narrative.

One should also consider whether the entire situation had been painted in too much black-and-white perspective. On the

---

68. See, for example, the decree published in Belorussia, cited in *Ekho Chernobylya*, No. 1, April 1990.

one hand, there is portrayed the supporters of glasnost and perestroika: journalists, those concerned about ecology, local doctors, concerned scientists and the general public. On the other remain what Shcherbak and others have described as the remnants of the Brezhnev bureaucratic system: giant Moscow-based ministries anxious more about plans than reality, concerned only to show that the effects of Chernobyl will not be serious in the long term, and to claim with irritating monotony that the reality must be left in the hands of "experts," all the more irritating when one might have considered that one of the abilities of a true expert in any field is the communication of the essence of his or her knowledge to the community at large.

The above represents part of the picture, perhaps even the major part. But it should be acknowledged that the position of a scientist like Bebeshko or Los is far from easy. They are not in a position to monitor every single sickness or feared sickness in Narodychi. To do so they would have to live in this area on a daily basis. Moreover, there have been some cases of irresponsible journalism and it would be remarkable if this were otherwise. Third, the role of Western authorities in this mounting crisis has often been overlooked as has the somewhat unrealistic perceptions of the infallibility of Western science in the Soviet Union. Thus if one is to carry a point in Soviet scientific society, the surest way to do so is to acquire some sort of confirmation from a scientist or institution in the West—witness the debate on the thirty-five rem radiation exposure limit.

Often overlooked is the attitude of the Western public generally to its own scientific community, its similar doubts and its prejudices. The deception over Chernobyl is comparable to that of the British government over the nuclear accident at Windscale in 1957, the effects of which were concealed for thirty years under the official secrecy act. They were concealed, it transpires, because the British did not wish their US allies to believe that they were incompetent in the areas of nuclear weapons manufacture. In the Soviet Union, there were mechanisms in place to hide the effects of disasters that were equally effective. What had been suspected by very few, however, was that in the Gorbachev period, these mechanisms were still in place.

Despite a one-sided approach and occasional reports that may have exaggerated—and perhaps exacerbated—the predicament, the Ukrainian media and elected politicians, the Rukh, Zelenyi svit, the Ukrainian Helsinki Union and other groups saw Chernobyl as the real turning point in Soviet society. The head of the Kiev branch of the Rukh, Volodymyr Yavorivsky, devoted most of a late-1989 Washington trip to the question of Chernobyl, an emphasis that surprised some of his listeners who may have expected more stress on the events of the recent Rukh Congress of September 1989. Chernobyl inspired Ukrainians to look at a number of issues more carefully, from the number of nuclear power plants being built in the republic to their environmental situation generally. It nurtured a lasting distrust for officials representing the leadership of the Communist Party of Ukraine, and for official accounts of what had happened at Chernobyl. The first and one of the most significant consequences was the development of a popular ecological movement in Ukraine.

4 | ECOLOGY
Irrigation and Nuclear Power Projects

THE ALARM GENERATED BY the Chernobyl fallout has heightened Ukrainian awareness toward ecological issues. While the impetus for this concern initially came from Moscow, the Ukrainian aspects of the Soviet ecology movement represent one of the most serious issues of today. It is impossible to discuss the economic future of Ukraine or the possibility of an economically sovereign Ukraine without taking into account environmental issues. Ecology has become a populist topic, one that raises emotions and frustrations and, moreover, one that encompasses a broad spectrum of the population. In essence it concerns the present and future existence of Ukrainians and whether they will live in a healthy or polluted climate. The problems that have been elucidated predate Mikhail Gorbachev, and many also predate the days of Leonid Brezhnev, who is widely cited as the main culprit for expanding industry so carelessly throughout the republic in the 1960s and 1970s.

In the Gorbachev period, the Ukrainian public acquired the chance of using newspapers as a forum for discussing the environment. For the first time in the Soviet period, the average citizen had a voice. Traditionally, whenever a reader sent a complaint about an environmental issue to a newspaper he generally received a haughty response from an official of the appropriate ministry; now, commonly, the entire editorial boards of newspapers became involved in discussions concerning ecological issues. Eventually, the number of letters on such topics began to fill pages and pages of newspapers for months at a time. By 1987, the question was arising of creating an informal ecological association that could act as a pressure group in situations where central planning had led to obvious environmental damage in a given area.

We will examine here several aspects of the ecological question in Ukraine and illustrate how the public has gradually strengthened its resolve and influence over living conditions. By the end of 1989, it was no longer possible for powerful all-Union ministries based in Moscow to dictate industrial expansion and planning. Scientific commissions, as in Narodychi, were also called upon to investigate potentially dangerous industrial enterprises, though their results were usually forwarded directly to the USSR Council of Ministers for a final decision. To provide a chronological account of the growth of environmental concerns in Ukraine, the first topic concerns land improvement and water economy, an area in which grandiose plans were already well in place by the time Gorbachev became General Secretary of the CC CPSU.

IRRIGATION SCHEMES ON THE EVE OF THE GORBACHEV PERIOD

Irrigation schemes in Ukraine have made use of the republic's natural river systems since the Stalin period, and more intensively since 1966. In May of that year, a campaign to introduce large-scale irrigation of arid lands was launched at a special plenary session of the CPSU Central Committee. Ukraine, along with the Baltic republics and Belorussia, and the Nonchernozem zone of the Russian republic, was part of an agricultural reform, launched by Brezhnev and Kosygin that anticipated a policy of raising the yields of grain on existing agricultural land through irrigation and drainage of fields. In the Ukrainian SSR, there were five basic projects under construction and relatively well advanced by the early 1980s, all of which were encompassed by the Brezhnev Food Program.

In 1966, construction had begun on the Lower Dniester irrigation system. Canals were to be constructed that would connect the Dniester to the Danube. The program did not keep to schedule,[1] but in 1973 it received a further boost when the salt-water

1. Radio Moscow, September 10, 1971.

Lake Sasyk was employed as a natural reservoir to irrigate arid zones in the area of the Black Sea.[2] The lake was subsequently dammed, drained and filled with water from the Danube by means of pumping stations established on its shores. In the mid-1970s, canals conveying the water to the Dniester reportedly irrigated some 150,000 hectares of farmland.[3] A second stage of this plan to utilize Danube waters to irrigate the arid lands of Ukraine was the Danube-Dnipro Canal, which was still at the planning stage in 1974. The original plan was to irrigate an area of 1.3 million hectares, which would represent 2.7 percent of the total arable land of Ukraine.[4]

The largest irrigation system in Ukraine at this time was called Kakhivska, and was expanded following the creation of a Kakhivska hydroelectric station in 1956. Officials boasted that this was the largest irrigation system in Europe at that time. Its main canal was 100 meters wide, 8.5 meters deep and 130 kilometers long.[5] Within this system was created a North Crimean Canal, the first stage of which was completed by 1975, and the second during the 1981–85 five-year plan. The canal's purpose was to irrigate farmland in the Crimean and Kherson Oblasts.[6] In the early 1980s, yet another canal was constructed to irrigate land in the oblasts of Dnipropetrovsk, Kharkiv, Donetsk and Voroshilovhrad. It was called the Dnipro-Donbass Canal, and contained within its system a smaller canal—the Pryazhivska—which was 260 kilometers in length and crossed into the Zaporizhzhya and Donetsk oblasts.[7]

These expansive schemes were typical of the Brezhnev era, a time when bureaucratic planning did not have to deal with public opinion or, in many cases, detailed scientific analysis of the likely consequences of such schemes beforehand. In October 1984, they received a further boost at a CPSU Central Committee Plenum under the Chernenko administration, at which time agriculture was under the control of Mikhail Gorbachev. It is somewhat ironic that he should be the architect of the future policy of

2. *Ibid.*, October 21, 1973.
3. *Ibid.*, August 13, 1978.
4. *Ibid.*, May 30, 1984.
5. *Sil's'ki visti,* June 6, 1976.
6. Radio Moscow, May 30, 1984.
7. Radio Kiev, August 26 and November 18, 1974.

glasnost that was to turn the public so dramatically and resolutely against them. However, even when he was in charge of agricultural affairs, he was known to be an opponent of such schemes. This plenum laid the groundwork for land improvement until the year 2000. In Ukraine, the plan was to increase the area of irrigated land from the 1984 total of 2.2 million hectares to between four and 4.2 million hectares, and that of drained land in the oblasts of Volyn, Rivne, Zhytomyr and Kiev from 2.6 to 3.9–4 million hectares.[8]

At this time, Ukraine was only part of some extremely ambitious schemes for the diversion of major rivers for agricultural purposes. Clearly the most significant involved a plan to reverse part of the flow of Siberian rivers to Kazakhstan and Central Asia. In September 1985, six months into the Gorbachev era, one Soviet source claimed that the discussion between proponents and opponents of the scheme was over, since it had been established that the plan would not be harmful to nature. The total length of the created canal was ultimately to be 2,230 kilometers, with a width of between 120 and 170 meters, and with an average depth of twelve meters.[9] Despite official optimism, this grandiose scheme was subjected to a barrage of opposition, and in turn such opposition helped to generate an increasing concern for the environment in the Soviet Union. One of the most vehement critics was the Soviet scientist Fadei Shipunov, who was to be a featured speaker at the first major ecological demonstration in Ukraine almost four years later.[10]

Land improvement in Ukraine has been an almost exclusively Soviet feature. Before the Soviet period, only about 430,000 hectares of land were subjected to drainage. In the 1930s, however, this figure was almost doubled, and by the mid-1980s, the figure was 2.9 million hectares, predominantly in the marshlands of western and northwestern Ukraine: in the oblasts of Lviv, Rivne, Zhytomyr, Volyn, Chernihiv and Transcarpathia. Large-scale marsh drainage has been attempted for years, but without much success in the Ukrainian Polissya, along the border with the Belorussian republic. Today this region encompasses some five

8. *Radyans'ka Ukraina*, November 3, 1984.
9. *Sotsialisticheskaya industriya*, September 4, 1985.
10. *Nash sovremennik*, No. 2, 1985.

million hectares of swampy and low-productivity lands that have contributed little to Ukrainian agriculture. The main area that has received water from irrigation canals is in the south and east of the republic: in the Kherson, Crimea, Odessa, Dnipropetrovsk, Donetsk, Zaporizhzhya and Mykolaiv oblasts.[11]

Following the 1984 CC CPSU Plenum, it might have been supposed that with the passing of Chernenko and the advent of Gorbachev, that the bold yet irresponsible plans for land improvement and irrigation that made up a part of the Brezhnev Food Program would have been cast aside. In fact, such plans have continued to have their supporters. In mid-1987, for example, an important article was published on the subject in the republic's economic journal, which represented the official viewpoint of Ukrainian planners at this time. Written by three authors, M. Lysenko, V. Hannalo and K. Prymak, the essay declared that the amelioration of land was one of the key directions in the intensification of agriculture, but improved lands in Ukraine—which made up 13 percent of the total agricultural area—were much smaller in proportion than those in similar agricultural regions of Eastern Europe, such as Bulgaria, with 31.2 percent, Romania 23.7 percent and East Germany 19 percent of cultivated land.[12]

Moreover, in Ukraine, the authors pointed out, two-thirds of the territory lacked conditions for the development of sustained agricultural production. In the period 1967–87, there had been eight years of acute drought, or approximately one dry year in every three. In the years when there was an adequate rainfall, the gross collection of grain in Ukraine had reached almost 50 million tons, but the drought reduced this total by 20–30 percent, particularly affecting crops such as potatoes, vegetables, fruits, grapes, technical (tobacco, sugar beets) and feed crops. The latter was having a particularly adverse effect, they claimed, on livestock breeding, which was taking up to five years to recover from losses during drought years. It has also been pointed out in another source that the amount of fresh water per head of population in Ukraine is nineteen times less than the all-Union average.[13] However, land improvement areas had risen dramatically in the republic since 1965, from 1.9 million to 5.5 million hectares.

11. *Tvarynnytstvo Ukrainy*, No. 10, October 1985.
12. *Ekonomika Radyans'koi Ukrainy*, No. 6, June 1987.
13. *Znannya ta pratsya*, April 1987, p. 6.

Within this latter total, the area of irrigated land had increased from 0.5 to 2.5 million hectares, whereas the amount of drained land rose from 1.4 to three million hectares. The driest region was the steppelands, where more than 80 percent of the land had been subjected to improvement techniques, while 75 percent of the land that had been drained was located in the northern district of Polissya. It was essential in the authors' view that these trends be maintained, so that by the end of the century the area of improved land would total more than eight million hectares. Already, in their view, such improvement upon nature was paying dividends: the 13 percent agricultural area that had been watered or drained accounted for 21 percent of total agricultural output on republican collective farms. In addition, the improved lands produced in 1985 12 percent of Ukrainian grain output, almost 60 percent of vegetables, more than 20 percent of the potato crop, 35 percent of flax and hemp, and 25 percent of feed crops.

The most significant indicator of the role of improved lands in agriculture, however, was to be found in the steppe zone of Dnipropetrovsk, Kherson, the Crimea and Donetsk. In the Crimean peninsula, irrigated lands which made up 19 percent of the total agricultural land produced more than 50 percent of the total output, including 60 percent of the feed crops and virtually all vegetables. Nevertheless, the authors acknowledged that improved lands had yet to fulfill all the functions that the planners desired. The harvests received were not meeting costs nor living up to their potential, so the primary task was to procure the more effective usage of such lands: in short, despite the alleged importance of such methods, they had not yet proven worthwhile from the economic perspective.

By 1988, however, Ukrainian specialists were becoming increasingly concerned about the effect of schemes that involved the diversion of rivers, the building of canals and general tampering with nature. One problem was the excessive amount of water that was being used, and much of it could not be recirculated. Between the Eighth Five-Year Plan (1966–70) and the Eleventh Five-Year Plan (1981–85), the volume of water consumption in the republic doubled and rivers were beginning to dry up. Enterprises and associations were ordered to reduce their water consumption by at least 15 percent by the year 1990.[14] The question of economizing on water usage was also studied by experts from the Ukrainian Academy of Sciences, and specifically by S.I. Dorohuntsov, a Candidate of Economics, and the Chairman of the Council for the Study of Productive Forces in Ukraine.

Dorohuntsov noted in a report to the Presidium of the Ukrainian Academy of Sciences that in a situation in which the Ukrainian economy was insufficiently provided with water, there had arisen a "negative tendency" to increase water consumption. The amount of water used per unit of production in Ukraine was reportedly much higher than in capitalist countries. The basic problem was that the water economy had been developed according to an extensive program, and had to be switched to an intensive one. Rather than concentrate on highly ambitious schemes, Dorohuntsov maintained, it should be possible to make better use of underground and sea waters. If one transfers the path of the natural water systems by building huge canals, he argued, then inevitably a considerable quantity of that water is lost and the damage to the ecology is also significant. In particular, the scientist took aim at the main Ukrainian construction project in the sphere of water economy: the Danube-Dnipro Canal.[15]

THE DANUBE-DNIPRO CANAL

We noted above that the Danube-Dnipro Canal constituted the second stage of plans to divert Danube waters for Ukrainian agriculture. The plan was approved by the USSR Ministry of the Water Economy, but it had its origins in the period before the Second World War. Its official elaboration was carried out by the Irrigation Management Institute of the Ukrainian Academy of Sciences. By 1961, a scheme was in place, approved by members of Comecon, and in 1963, the Soviet Gosplan confirmed that the Soviet Union needed some 30–35 billion cubic meters of water from the Danube to irrigate land in Moldavia, Ukraine and the RSFSR. The Danube-Dnipro Canal was approved by an expert commission of this same State Planning authority. The canal was

14. *Pid praporom leninizmu*, No. 9, May 1988, p. 8.
15. *Visnyk Akademii nauk Ukrains'koi RSR*, No. 3, 1988, pp. 9–10.

to be some 300 kilometers in length, and was to be located in the oblasts of Odessa, Kherson and Mykolaiv, running from the mouth of the Danube to the estuary of the Dnipro and Buh rivers in south Ukraine.

This megaproject also included the North Crimean Canal, noted earlier (though the idea for this canal dates back to the mid-nineteenth century), and the construction of a huge hydro-electric station on the Dnipro-Buh estuary, to which a large dam would be linked. This dam, in turn, was to separate the Dnipro River from the Black Sea. In the first stage of the Danube-Dnipro Canal, which was to have been completed by the end of the century, the plan was to collect seven billion cubic meters of water, and ultimately to raise this total to twenty-five billion cubic meters, 80–85 percent of which would be used to irrigate the arid lands of the Ukrainian steppe.[16] Those who have supported the building of the canal have argued that its plans were very similar in scope to plans to divert Danube waters in countries such as the Federal Republic of Germany, Hungary, Czechoslovakia and Poland.

One such advocate was B. Strelets, Deputy Minister of Land Improvement of the Ukrainian SSR. He pointed out that the plans for the Canal had already been amended in order to preserve the natural environment. The Danube waters, which had been criticized for having a high mineral content, were in his view cleaner than the Dniester and South Buh rivers. Moreover, for the most part, the "load" being placed on the Danube in Ukraine was far from heavy. In 1987, some five million hectares of land were being irrigated with its usage, while the future figure would be seventeen million.[17]

In the Gorbachev period, the Ukrainian public took an increased interest in the Danube-Dnipro Canal. It became the first major ecological issue after Chernobyl in which a virtual popular movement began to take effect. One of the more prominent Kiev newspapers, *Robitnycha hazeta* (Workers' Gazette), which is published in both Russian and Ukrainian editions, lent its columns specifically to the issue in 1987. While the editor of this newspaper, Mykola Shybyk, a member of the USSR Supreme So-

---

16. *Radyans'ka Ukraina*, August 18, 1988.
17. *Robitnycha hazeta*, August 21, 1987.

viet, has a reputation for being a conservative—that is, as a man who has not supported the program of glasnost and perestroika—he did not evidently impose any restrictions on his editorial board insofar as the publication of letters hostile to the Canal was concerned.

The Ukrainian public was clearly upset by the salination of Lake Sasyk during the construction of the North Crimean Canal. Water had filtered through the walls of this canal so that limestone caves in the Crimea had become flooded with chemically destructive water.[18] At the request of the Commission for the Study of Productive Forces and Natural Resources, affiliated with the Ukrainian Academy of Sciences, a group of scientists traveled along the planned route of the canal in the summer of 1988, mainly in the area of Lake Sasyk. They noted the "gross ignorance" on the part of the planners, and remarked that Lake Sasyk had yet to be desalinated. Sulfuric acid was being dumped into the canals in an attempt to purify the water. However, the soil in the region was said to be dying. At the bottom of the Dnipro-Buh estuary, moreover, a layer of saline silt was visible, testifying to the damage caused by irrigation schemes.

By the late summer of 1987, the Ukrainian State Nature Committee had become involved in a reinvestigation of the ecological and economic viability of the Danube-Dnipro Canal. The Chairman of this committee at that time, D. Protsenko, commented that the plans for the Canal had never been properly approved, and that another Expert Commission of Gosplan had the task of submitting revisions to the original plan by the year 1990. Protsenko noted that as a result of the scale and complexity of the project, the Presidium of the Ukrainian Academy of Sciences had created an Interdepartmental Commission for Economic and Ecological Problems in the Construction of the Danube-Dnipro Water Economy Complex. The commission has the job of studying the possible consequences of the plans for the Canal.[19]

Protsenko and the Ukrainian State Nature Committee took a mildly critical attitude toward the project, which, as will be seen, simply failed to measure up to the growing public alarm. In March 1987, he stated, the State Committee had convoked a

18. *Ukraine*, No. 7, 1988, p. 14.
19. *Robitnycha hazeta*, September 11, 1987.

meeting of experts to examine the plan to build the Dnipro-Buh hydroelectric station, and concluded that there were problems with it. Noting that the CC CPSU and USSR government had cancelled the plan to divert the flow of northern and Siberian rivers, it was suggested that alternative variants should be submitted for providing water to the southern regions of Ukraine. In economic terms, it was considered opportune to make the transition from extensive to intensive programs of development. The Irrigation Management Institute was to be advised that it must determine the ecological effects of damming the estuary, and the negative effects this might have on the resort areas of the Black Sea. In addition, the Committee recommended that important work should be undertaken to raise the water expanse of small rivers, reconstruct the Kakhivska and Kremenchuh reservoirs and develop plans for the more rational use of underground waters.

A more critical analysis at this time was offered by V. Motsarenko, a scientist with the Odessa branch of the Institute of Economics, at the Ukrainian Academy of Sciences. He made reference to what he termed the "fiasco" at Lake Sasyk, a manmade lake that began as an experimental base for the desalination of sea water. Not only had this lake become salinated because of "miscalculations" on the part of planners, but the waters of the Sasyk estuary, harnessed for irrigation needs in the Crimea, had caused severe deterioration of the soil quality in the irrigated region. An impenetrable crust, dense and salinated, had formed at the soil surface, deposited there by the irrigation water itself.[20]

To offset this damage, a bypass canal had been built at the southern end of the reservoir, doubling the cost of the overall structure. But problems remained, partly because of a failure to resolve the problem of the salination of the reservoir bed, and the lateral underflow of underground salt water. Motsarenko pointed out that the harvest in the fields under irrigation in 1987 was 2–2.5 times lower than had been projected. The plan anticipated a recoupment of the investment over a period of nine years—itself considerably longer than the usual recoupment period[21]—but bearing in mind the costs of desalination, the reduc-

20. *Ibid.*, September 25, 1987,
21. An article by a Ukrainian land reclamation official, published in 1985, maintained that the "correct" time period for the recoupment of investment into

tion of the soil's fertility and the resultant low harvest, it seemed that the capital investment had hardly been recouped at all.

According to Motsarenko, there were far more profitable ways to improve agricultural land. Irrigation, he stated, was but one of forty types of land improvement, but in an average year, six billion rubles was being expended on it, compared to only 0.2 billion rubles in all other types of reclamation. On average, it was costing between 4,000 and 7,000 rubles to irrigate one hectare of land, compared to only 200 to 300 rubles for field-protection afforestation. But whereas the latter method yielded some 450 rubles annually in profits, more than covering the amount of investment, irrigation yields were not even covering costs. Moreover, while irrigation may have increased crop yields, the amount of the increase was significantly less than on lands that had been improved by other means. And already, 380,000 hectares of secondary agricultural land had been salinated in Kherson Oblast, and the mouth of the Dniester River had been destroyed. How under such circumstances, he wondered, could planners move ahead with the Danube-Dnipro Canal?

At this same time, the Ukrainian Academy of Sciences was given the task of making a recommendation about the viability of the Danube-Dnipro Canal, which was to be submitted to the Ukrainian government by December 1988. This was to be one of many controversial schemes subjected to a new scientific inquiry. Meanwhile, the public was writing to *Robitnycha hazeta* on the same subject and its verdict was overwhelmingly negative. G. Kostak, a biologist from Volyn Oblast in western Ukraine, commented that the planners of the Danube-Dnipro complex were dilettantes when it came to ecology. He maintained that he had visited eleven Ukrainian oblasts and observed several small rivers that had dried up. Would it not be better, he asked, to "rescue" these small rivers rather than draw up grandiose projects? The next generation would not thank the planners for their many thoughtless interferences with nature, he concluded.[22]

V. Bocharov, a raion prosecutor, recalled the "sad lessons" of

---

land improvement was three to four years for irrigated land, and five to six years for drained land. See *Tvarynnystvo Ukrainy*, No. 10, October 1985, p. 20. In 1979, it was reported that seven to eight years were normally required for the recoupment of investment costs. See *Pravda*, June 4, 1979.

22. *Robitnycha hazeta*, September 11, 1987.

Lake Baikal, the Volga, the Sevan estuary and then demanded an end to the Dnipro-Buh hydroelectric station. It had never been officially approved, he stated, but over 100 million rubles had already been spent on the project. Enormous sums had been expended on the alluvium damming of the construction site: "No one thinks about ecology. How long are we to continue these depraved practices?" he lamented.[23] B. Vorontsov of Kiev Oblast wrote that there was no water shortage in Ukraine, but rather, the authorities had wasted existing supplies. Drinking water was being used for washing laundry, the streets and automobiles. In his view, it was to be hoped that Ukraine would be fortunate enough to "escape" from this canal, just as other regions had avoided the project to divert the northern rivers.[24]

The general consensus was that the Danube-Dnipro Canal was an example of "gigantomania" and that such projects had been elaborated during the "period of stagnation," associated with the time when Leonid Brezhnev was General Secretary. Generally speaking, the public did not yet take this one stage further to assert that behind the plans were ministries that had become dictatorial and whose officials, for the most part, had not set foot in Ukraine, let alone made a detailed analysis of the effects of their schemes in nature. The list of organizations that opposed the plan had grown considerably by 1988 to the stage where the planners found themselves in a distinct minority. The opposition included the Ukrainian Academy of Sciences, the Council for the Study of Productive Forces in Ukraine and the Committee for the Protection of Nature of the republic. It was to provide one of the foundation stones for the initiation and development of the ecological association, Zelenyi svit (Green World).

However, the discussion was not completely one-sided. From time to time, officials from the Ministry of Water Economy would defend their position. Also, on occasion, representatives from areas that were to receive the irrigated water would make a stand. One such instance took place in January 1989 when the Canal was defended by A. Roshchupin, the Chairman of the Crimean Oblast Executive Committee. In a bitter article, Roshchupin noted that in the past, deaths had resulted from water

23. *Ibid.*, September 18, 1987.
24. *Ibid.*

shortages in the Crimea, and the amount of rainfall on the peninsula had never exceeded 350 millimeters, even in favorable years.
The first two stages of the North Crimean Canal had solved this
problem, he added. He stated that they had also resulted in an
improved demographic situation, inasmuch as the rural population of the oblast had risen by over 200,000 as a result of the
canal's operation.

As for problems associated with the North Crimean Canal,
Roshchupin was prepared to acknowledge that its lack of adequate lining did have a negative impact on the environment.
Without going into the details of the effects of the original poor
construction, however, he asserted that over the past fifteen years
the canals had been provided with concrete walls and lined with
polyethylene. He also maintained that new pumping stations considerably reduced the loss of water during filtration. The introduction of new water sprinklers and careful crop rotation had, in
his view, enhanced usage of irrigated land. Claims that irrigation
led to soil degradation, on the other hand, were said to be unfounded and subjective. Those who had attacked the Ukrainian
Ministry for Land Improvement and Water Economy, as a rule,
were "incompetent hotheads" whose arguments were based on
subjective emotions, and not a healthy realism.[25]

In the final analysis, the discussion centered on the economic
needs of the Ukrainian republic. It had often been pointed out
that Ukraine, with less than 3 percent of Soviet territory, accounts for 25 percent of Soviet agricultural production. In addition, it possesses a disproportionate amount of heavy, water-
consuming industries, such as ferrous metallurgy, machine building, nuclear power plants and chemical works. I.I. Lukinov, the
Vice-President of the Ukrainian Academy of Sciences, pointed out
that water consumption in Ukrainian industries was 2.5–4 times
higher than in the developed capitalist countries of Europe.[26] Each
1,000 megawatt nuclear reactor consumed thirty million cubic
meters of water per year. Water consumption was said to be
doubling every fifteen to twenty years. The solution, then, appeared to be to reduce the output of certain industries that consumed excessive amounts of water, while cutting down on water
consumption generally.

25. *Pravda Ukrainy,* January 26, 1989.
26. *Robitnycha hazeta,* March 20, 1988.

With such goals in mind, the Chairman of the Ukrainian State Planning Commission, V. Fokin, announced in late November 1988 that the main goals of the Danube-Dnipro Canal—specifically to cut the Dnipro River off from the Black Sea by building a huge dam—had been abandoned permanently.[27] The work on surveying and constructing the waterway was to be halted by 1989, and the preparation work on the Dnipro-Buh dam was also to be halted. Over the next fifteen years, reconstruction work was planned on the Kremenchuk and Kakhivska reservoirs. This decision represented an important victory for those who had opposed the plan. It can be contrasted with the failure to shut down the Chernobyl nuclear power plant after the major accident. It also signified that there were to be serious structural changes in the Ukrainian economy. By the year 2005, it was announced, there was to be a substantial reduction in the output of commodities that used up large amounts of water: cast iron output was to be cut down by 30 percent; steel by 29 percent, and coking coal by 33 percent.[28]

The halt on the Canal's development has been far from total, however. Work on the North Crimean complex and around Lake Sasyk has continued. There have also been constant criticisms about the failure to supply some towns in southern Ukraine with an adequate water supply from existing irrigation projects. Nevertheless, the Canal was the first megaproject to be stopped in Ukraine, and it paved the way for ecological populism to become part of public life. Henceforth, any project that appeared to threaten nature was going to meet with opposition. In addition, the average citizen, through the debates in newspapers, and on television and radio, was becoming more ecologically conscious. The fact that such consciousness had come belatedly to Ukrainians, as might be expected, was reflected in a higher degree of fanaticism and intolerance toward such projects. It is characteristic of the debates on ecology in Ukraine that they should be painted in black and white colors, omitting any shade of gray. On several occasions, the editors of newspapers would simply add their weight to the protests. In this way, some of the major correspon-

27. *Robitnycha hazeta*, November 27, 1988.
28. *Izvestiya*, November 26, 1988. See also *News From Ukraine*, No. 3, 1989.

dents for Kiev newspapers were to become prominent in the ecological movement.

Another series of protests were conducted parallel to the campaign to halt the Danube-Dnipro Canal that were also to have a major impact on the Ukrainian economy, namely those against the construction of nuclear power plants. Such major consumers of water had been constructed ostensibly because they were ecologically clean sources of power. After Chernobyl, however, it became impossible for advocates to make such claims. Moreover, the fact that Ukraine had been designated as the center of the nuclear power program added oil to the fire that was spreading. The victory over the Canal also intensified the efforts to stop the building of new reactors on Ukrainian territory partly for the same reason: that the republic simply lacked sufficient supplies of water to meet the expansionist plans in nuclear energy.

PROTESTS AGAINST NUCLEAR POWER PLANTS

Nuclear power provided an excellent example of the way in which an old-style bureaucracy appeared to be still in place in the Soviet Union. Officials based in Moscow, at all-Union ministries, were still making decisions about nuclear power plants scattered across the Soviet Union, often without any previous discussion with local residents. In the past, it had not been customary for such plans to be discussed publicly. There was no open discussion on an issue such as nuclear power, partly because the industry had a key military dimension, and partly because it was felt to be self-evident that the country needed the electricity that would be generated, and without it would find it difficult to survive as a modern state. It is worth noting that this supposition is hardly unique to the Soviet Union and that even in the democratic West, discussion on the issue of nuclear power has often brought out the worst of two sides rather than reasoned debate.

In addition, Chernobyl was akin to an earthquake that contained a much greater aftershock. The number of victims of Chernobyl is rising much more quickly today than in 1986, and even the most partisan observers can hardly deny that the story, as presented officially by the Soviet authorities, was not a full one and in fact concealed most of the problems caused by the disaster. As a result, thousands of people became sick in areas that were not evacuated. Moreover, it is a far cry from academic institutions that may see debate on issues such as the strengths and weaknesses of the RBMK to the lonely villages of northern Ukraine or Belorussia that have to live with the consequences. And in Ukraine, it would be harsh to judge the response to Chernobyl purely in scholarly or scientific terms. It has seen an emotional outpouring of despair; it has seen the sometimes violent campaign against nuclear power plants generally. There is no faith in nuclear power in Ukraine, and yet it was here that the republic's economic future rested. A giant expansion program thus became the prime target of protesters from all walks of life.

It is well known that Ukraine had become the focal point of the Soviet nuclear power program and that at least eleven stations were planned with an ultimate total generating capacity of more than 42,000 megawatts. When all the proposed reactors were in operation, it was believed, nuclear power would account for 60 percent of electricity generation in the republic. The pros and cons of nuclear power have been widely debated in Ukraine, but in essence they have not been the catalyst to protest movements. Instead, one can put the situation more simply: Chernobyl poisoned the mind of the public against nuclear power irrevocably. Moreover, the industry had traditionally been operated in such a centralized and secretive fashion that it appeared to embody the worst features of the Brezhnev period bureaucracy, which after 1986 came under increasing attack in the Soviet Union. From both the ecological and political perspectives therefore, an attack on nuclear power stations became an integral part of glasnost and perestroika. If the process was more acute in Ukraine than elsewhere, the reasons for such extremes were manifest: Chernobyl; and the management of the existing Soviet nuclear energy program. We will look first at the general situation in the Soviet Union on this emotional question before turning specifically to the individual Ukrainian projects.

THE POST-CHERNOBYL NUCLEAR PROGRAM

A combination of public anxiety and official mishaps raised the controversy over nuclear power to new levels. To give just one example, in mid-November 1989, rumors spread across the heavily populated regions of Eastern Ukraine that an accident involving the release of radiation had occurred at the giant Zaporizhzhya station (5,000 megawatts in size at this time, based on five VVER reactors), the largest in the republic. Officials at the plant were quick to assure the public that no such accident had occurred. One spokesperson at Energodar, the city that houses plant operatives, maintained that such rumors were a result of "radiophobia," that had arisen ultimately from Chernobyl. Barely one week later, however, a fire broke out at the building site of the third reactor unit of the Khmelnytsky station in western Ukraine. The plant Director, V.G. Sapronov, assured the press that the fire had been contained, and had not spread to the operational first unit or the almost completed second unit.[29] Nevertheless, the reaction to the two events indicated the public's nervousness about nuclear power.

The fall of 1988 saw a series of public protests against the installation of nuclear reactors in various parts of the Soviet Union. In contrast with the earlier protests, these campaigns were massive in scale, involving thousands of people in protests, and signatures on petitions. There appeared to be two major concerns: first that new power stations being brought on stream and additional reactors at existing stations posed a serious threat to the environment; and, second, that the stations—and particularly those with graphite-moderated, water-cooled reactors (the RBMK-1000 and RBMK-1500)—were unsafe, a view on which there was a degree of consensus at almost all levels of Soviet scientific society. Between 1987 and the summer of 1988, protests had virtually halted a planned new nuclear power station in Armenia. An RBMK-1500 unit being built in Kostroma was quietly being converted to a VVER reactor unit. In October 1987, a planned combined nuclear power and heating station near Kiev was switched to a hydrocarbon-burning thermal electric station. In January 1988, the Krasnodar nuclear plant was abandoned after public protests with a loss of at least fourteen million rubles of investment.[30]

There had long been concern also with the existing VVER-440 reactor in Armenia. At a meeting of the International Atomic En-

29. *Molod' Ukrainy*, November 19, 1989.
30. See Elizabeth Fuller, "Armenian Authorities Appear to Yield to 'Ecological Lobby'," *Radio Liberty Research Bulletin*, RL 130/87, March 30, 1987; and *Pravda*, January 21, 1988.

ergy Agency in Vienna, a Soviet representative, Boris Semenov, informed journalists that there was some anxiety because the station was based on first-generation nuclear technology, and was located (like the Crimean station) in a region noted for its seismic activity.[31] By January 1989, it had been announced that the Armenian station would be shut down by March, an event precipitated by the Armenian earthquake. Similarly, in Belorussia, a nuclear-powered heat and current plant being built near the city of Minsk was halted early in September 1988 following public fears about the operation of such a station in a densely populated area. The station was replaced with a thermal electric station burning natural gas. The Minsk station was to have been located only about twenty-two miles from the city center, the proximity necessitated by the significant loss of power during transmission over larger distances.[32]

The situation was to become particularly complex in Lithuania, a republic which declared independence from the Soviet Union on March 11, 1990. At Ignalina, close to the junction of the Belorussian, Lithuanian and Latvian borders, a giant nuclear power plant stands, based on RBMK-1500 reactors. By the summer of 1988, two reactors were operational, but it had been announced that a third phase of the plant (which could have signified planned fifth and sixth reactors because they are usually built in twins) had been suspended pending a safety review. Less than one week later, unit two was shut down after a fire had broken out in the reactor control cables. By mid-September, the fledgling Initiative Group in Support of Perestroika in Lithuania had amassed 287,000 signatures on a petition demanding that the plant be investigated by an international commission. In addition, several thousand people held a demonstration near the plant in this same month, forming a human chain to express their anxiety about the station's safety.[33]

The fall of 1989 also saw protests against the proposed or continuing construction of several Soviet nuclear power plants. In late October, a five-hour meeting was held in the city of

31. *Radio Free Europe/Radio Liberty Special Report,* September 23, 1988.
32. *Izvestiya,* September 6, 1988.
33. *Izvestiya,* August 31, 1988; *Reuters,* September 14, 1988; *TASS,* September 17, 1988; and *Radio Free Europe/Radio Liberty Special Report,* September 23, 1988.

Neftekamsk to debate the wisdom of building the Bashkir nuclear power plant, only twenty miles from the city. The result was a resolution that construction of the station must be stopped, and a final decision was to be taken by the Soviet government. A similar situation was reported in the Tatar republic concerning the proposed construction of a nuclear power station there, where the atmosphere was said to be "boiling." There were at least three protest marches against the project in late September and early October. Finally, in September 1989, residents of Chelyabinsk made known their anxieties and anger, and demanded a referendum on the question of the South Urals nuclear power plant.[34]

The current Soviet nuclear program has seen several stages, commencing in the late 1950s and early 1960s. The two prototype industrial nuclear plants were at Beloyarsk and Novovoronezh. The former converted the military graphite type of reactor (RBMK) to the civilian industry, while the latter saw the development of the water-pressurized reactor (the VVER) of 440 megawatts capacity. The development of the VVER resulted in the construction of similar reactors at nuclear stations in Armenia, Kolsk and Rivne. In 1980, a fifth unit at Novovoronezh became the prototype of the VVER-1000 design that has become dominant today in the Soviet Union and the East European countries, and led directly to the first stages of building at the South Ukraine and Kalinin plants.[35]

The RBMK-1000 was first developed commercially at Leningrad, a station that reached its designated capacity of 4,000 megawatts in 1981, and led to the building of similar units at Kursk, Chernobyl and Smolensk, and eventually to the 1,500 megawatt unit in Lithuania. The expansive program became even more ambitious with the development of the so-called flowline process at Zaporizhzhya, designed by the "Atomelektroproekt" institute. The authorities have claimed that the flowline system (using huge prefabricated building blocks weighing up to 120 tons each) enabled a reduction of time and labor costs of at least 30 percent, and the building of the first unit at Zaporizhzhya was reportedly 25 percent cheaper than that of South Ukraine-1,

34. Radio Moscow, September 12, October 11, and October 26, 1989.
35. *Atomnaya nauka i tekhnika*, Moscow, 1987, pp. 46–48.

which has an identical reactor type. Reactors being built according to the RBMK design at Kursk, Chernobyl and Smolensk were switched to this system after 1984, as were VVER-1000 units at Balakovo, Rivne (unit 3), Crimea, Rostov and Khmelnytsky. The "success" at Zaporizhzhya encouraged planners to devise more ambitious programs to be fulfilled by the end of the century.

After Chernobyl, the official program remained expansive. In early 1989, Aleksandr Lapshin, Deputy Minister of Nuclear Energy, announced that Soviet nuclear energy capacity was to rise from 35,400 megawatts (January 1989) to 100,000 megawatts by the year 2000. At this time, there were said to be seventeen nuclear power stations in operation, with a total of forty-seven reactors, making up 13 percent of the country's total electricity generation.[36] The latter figure prompted the pro-nuclear power lobby in the Soviet Union to declare that the Soviet program is still in the process of catching up with more advanced Western nations, such as the United States and France. Thus in September 1988, F. Bromberg, a senior scientific worker at the Institute of World Economics and International Relations, pointed out that French nuclear power plants account for 72 percent of that country's electricity generation, while the percentage in West Germany and Japan is 25–30, and in the United States and Canada, 16–18 percent, all higher than the Soviet proportion.[37]

Basically, Bromberg pointed out, these foreign percentages were the result of the completion of existing programs. In countries such as Sweden, he indicated, nuclear power was being phased out at the behest of the population, and will cease to exist by the year 2010. But the USSR simply wishes to complete an existing program and thereby to attain the sort of percentage level of nuclear-generated electricity reached in other industrialized nations. The problem with this analysis is that it did not take into account the attempts to commission new constructions that would triple existing capacity if completed. At the start of 1987, twenty-six VVER-1000 reactors were at the construction stage at the following sites: Kalinin (2), Balakovo (3), the Tatar republic (4), Rostov (4), Bashkir (4), South Urals (2), Zaporizhzhya (1), Khmelnytsky (1), Rivne (1), South Ukraine (2),

---

36. *TASS,* January 29, 1989; and *Pravitel'stvennye vestnik,* No. 2, January 1989.
37. *Stroitel'naya gazeta,* September 28, 1988.

the Crimea (2), with the addition of four VVER AST-500 heat supply stations, of which two were being built at Gorky and two at Voronezh.

Among the numerous discussions in 1989 about the viability of nuclear power in the USSR in the current climate, one by Boris Kurkin encapsulated the existing problems most succinctly.[38] Kurkin noted that over the past fifteen years, the costs of building nuclear plants have risen by 400 percent, partly as a result of rising fuel prices, but mainly because of heavy expenses on the infrastructure that must accompany such a major building: residence building, road construction, social and cultural amenities. Long delays have occurred between the time a potential site is selected, and the time that building work actually begins. For example, there was a five-year delay at the Smolensk site, he pointed out, and a ten-year gap at Kalinin.

Kurkin estimated that on average in the Soviet Union, it took one worker to produce one megawatt of electric power, whereas only 0.2 workers were required in the United States. In addition, the shortage of advanced computer systems has meant that the number of staff at a Soviet nuclear plant is much higher than the world average, and the usual size of the town for plant operatives is 35–40,000 in population. Moreover, dismantling the plants after a 25–30 year lifespan should be a further factor in assessing costs of construction, Kurkin postulated. Existing sites were expected to contain radioactive materials for 50–100 years after the plant had completed its period of service, and these had to be preserved for the safety of the population. When a decision was being made to build a new nuclear plant, therefore, planners should, in his view, take into consideration five factors: direct capital investment; the protection of the surrounding environment; social and economic factors (including the building of the necessary infrastructure); the health and safety of the nearby population; and public opinion, that is, the desire or nondesire of the local population to have a nuclear plant built in the vicinity.

It has been this final category—public opinion—that has made life very difficult for the planners of the nuclear power program. However, for the past few years, and especially since the Chernobyl disaster, Soviet scientists have been attempting to build a

38. *Molod' Ukrainy*, February 15 and 19, 1989.

safer reactor. Some have maintained that the plans for building the so-called "fast breeder" reactors, which also have the advantage of using for fuel uranium-238, which is far more plentiful than the uranium-235 that is used as fuel by the majority of reactors, may represent a solution. Such reactors (designated as the BN series) are also said to have longer periods of operation than RBMKs and VVERs, lasting from thirty to fifty years. A BN-350 was first built in 1973 on the Mangyshlak peninsula of the Caspian Sea, and it has been developed further with the Beloyarsk BN-600 station in Sverdlovsk oblast.[39] In the summer of 1986, the Atommash factory in Volgodonsk, the main producer of the VVER, began work on a fast neutron reactor of 800 megawatts in capacity, and ultimately Soviet designers look toward a BN-2000. Such a station would be more economical than existing plants, with a longer lifespan and less of a drain on uranium supplies. Nevertheless, the construction costs of such a reactor are prohibitively high.

It is with this background in mind, with a very negative public attitude toward nuclear power, that one must view the problems and protests in Ukraine, which added fuel to the environmental movement. While the background to this movement in Ukraine dates back to late 1986, our focus will be on the period after the summer of 1987, and particularly on the period 1988–90.

ZAPORIZHZHYA

As noted, the Zaporizhzhya nuclear power plant is not only the largest, but also one of the most ambitious nuclear power complexes in the republic. Further, until the summer of 1988, it was notable for being the least criticized of all the Ukrainian projects. It has been, in short, a model plant. But in July 1988, the first public problems emerged, and, significantly, from within the plant rather than from the public. Fifteen construction and assembly workers wrote a letter to the editors of a major newspaper, which complained both of the timetable for the assembly of a new fifth 1,000 megawatt reactor, and asked whether such a reactor was necessary to the economy.[40] The latter question was

---

39. *Sotsialisticheskaya industriya,* August 16, 1988.
40. *Stroitel'naya gazeta,* July 15, 1988.

to become a regular one, because those concerned with the environment in Ukraine have maintained with consistency that much of the extra generating capacity being commissioned in the republic was for external usage, either elsewhere in the Soviet Union, or in Eastern European countries, such as Poland, Romania, Hungary and Bulgaria.

Although Zaporizhzhya has seen the commissioning of reactors practically at annual intervals through its flowline system, it has been using an inordinate amount of labor—some 12,000 workers are thus employed on a full-time basis according to the letter. Moreover, delays in the supply of equipment to the site at Energodar were first highlighted in October, 1986, when First Party Secretary, Volodymyr Shcherbytsky, paid a visit to the site. At that time, R. Khenokha, the director of the construction department, stated that, following the Chernobyl disaster, his staff were of the opinion that it was not feasible to continue to adhere to the rigid construction schedules because the supply system was so weak.[41] The problem with the supply of equipment was equally severe two years later. The letter to the newspaper noted that items not delivered to the station included 16,371 pieces of equipment and electrical apparatus, forty-seven tons of piping, and 530 kilometers of cable. The key questions were whether the lack of these items would affect the timetable for commissioning new capacity and, if not, whether the new reactor which was scheduled for December 1988 (Zaporizhzhya-5), would be operating with inferior or defective equipment. The newspaper put these questions to Deputy Minister of Nuclear Energy, Evgenii Reshetnikov.

Reshetnikov laid the blame for the disruption of supplies squarely on the shoulders of various ministries. In particular, the Ministry of Chemical Machine Building, the Ministry of Heavy Machine Building, and the Atommash plant, he stated, were most responsible for the delays. However, he added, all the nuclear reactor units currently under construction were suffering similar problems. Reshetnikov was also asked a number of questions about the feasibility of bringing so many reactors on stream in one year (the schedule cited not only Zaporizhzhya, but also Smolensk, South Ukraine, Rostov and Balakovo). In response, he

---

41. *Izvestiya*, October 12, 1986.

noted that the plans of the Ministry of Atomic Energy called for a total of seven new reactors in 1989 and eight in 1990, all 1,000 megawatt units, so that by comparison, the 1988 program was relatively modest.

On a personal note, nonetheless, Reshetnikov maintained that while there was an adequate supply of major equipment to ensure the startup of five reactors per year, the shortage of auxiliary equipment, as at Zaporizhzhya, reduced the practical number to three. He declared that some of the equipment being manufactured was "junk," and that it was time to stop "entertaining illusions of impossible plans." The significance of this event in our context is that it indicated a discontented workforce, and that despite official confidence about nuclear energy, those on the worksite itself were far from happy, and had begun to question the viability of the overall program. If difficulties were manifesting themselves at Zaporizhzhya, however, they paled in comparison with those in the Crimea, where a reactor on the Kerch peninsula was scheduled for completion by 1985 and, while delayed, was clearly ready for service three years later.

CRIMEA

The Crimean nuclear power plant had long remained one of the most unpopular projects in Ukraine. Located on the Kerch peninsula, it was scheduled to come into service in the 1981–85 five-year plan, but was then delayed. Local Crimeans, who were incensed by the industrialization of their once picturesque peninsula, and who were embroiled in the question of whether the Crimean Tatars, deported from their homeland by Stalin during the Second World War, should be permitted to return, saw the nuclear plant as yet another example of the workings of a centralized and uncaring bureaucracy. Further, emotions aside, there appeared to be grounds for concern regarding the area in which the station was being built, which was known to be one of acute seismic activity. In September 1988, therefore, the Soviet government appointed a high-level commission from the USSR Academy of Sciences, under the leadership of Vice-President, Evgenii Velikhov—a well known proponent of nuclear energy—to investigate the Crimean nuclear plant. The results of the commission's work were published in the Kiev press in November of this year,

1(a) (above). Chernobyl: No 4 reactor as viewed by helicopter shortly after the explosion. (A. Salmygin 'Kombinat' Association, Chernobyl)

1(b) (right). A tree warped by radiation from the nearby Chernobyl station (A. Salmygin, 'Kombinat' Association, Chernobyl)

2. The end of a human chain linking Lviv with Kiev on the eve of Ukrainian Independence Day,

3. A pre-election meeting, February 11, 1990, October Revolution Square, Kiev.

4. A silent protest in Kiev against the refusal to register Green World candidates. The sign reads 'Who registered the CPSU and when?' February 1990. (photo Oleh Kulchytsky).

. Anti-Chernobyl rally in Kiev. To the left of the microphone is Volodymyr
Yavorivsky, people's deputy and chairman of the Kiev branch of Narodnyi Rukh.

6. Anti-Chernobyl protest in St. Sophia Square, Kiev, April 26, 1990. The sign reads 'Moscow! Give us money for the liquidation of the accident.'

7. Protests outside the Ukrainian government building on the fourth anniversary of Chernobyl, April 26, 1990.

8. Green World members protest against the government's refusal to register their party for the Ukrainian elections, spring 1990. The sign reads 'We demand elections under United Nations control! Are we worse than Namibians? (photo Oleh Kulchytsky)

9. Deputies in parliament from the Lviv region applaud the declaration of Ukrainian sovereignty, July 16, 1990.

and were analyzed by a prominent environmentalist, A. Hlazovyi.[42]

Hlazovyi noted the changing attitude toward nuclear power in the Soviet Union over the previous twelve years. At one time, Anatolii Aleksandrov, then president of the USSR Academy of Sciences, had argued that there was as much chance of an accident at a nuclear plant as there was of the Earth being struck by a huge meteorite. But views had changed because economic activities were supposed to be linked to ecological concerns. However, this did not appear to be the case in the Crimea. Hlazovyi cited the example of Evgeniya Lvova, a candidate of geological sciences from Simferopol, who had worked in the Crimea for almost forty years and was well acquainted with the geology of the region. When the idea of building the station was first broached in 1974, she strongly opposed it and wrote a letter expressing concern about the plan to the authorities. Soviet nuclear energy officials responded that her fears were groundless. Her voice, therefore, "was not heard," and construction of the Crimean plant began.

In the fall of 1988, the Crimean Komsomol newspaper published a questionnaire asking whether the Crimean plant was needed. It received almost 10,000 responses, containing over 30,000 signatures. Of the 30,000, only two respondents were in favor of building the station. Adding these to the results of other newspaper surveys that had been published, stated Hlazovyi, there were now some 250,000 signatures against the station's operation. He took issue with the views of Aleksandr Lapshin, Deputy Minister of Nuclear Energy of the USSR, on this matter, following an interview with the minister in the newspaper *Krymska pravda* on October 27, 1988. According to Lapshin, the Soviet Union was not alone among the world's developed countries in continuing to build nuclear reactors after Chernobyl: the Japanese and French were expanding their facilities. But Hlazovyi maintained that the Japanese reactors were far superior in quality to those in the Soviet Union, while France was the unique example in the world of an advanced country mainly reliant on nuclear power for its electricity generation.

Lapshin had also played down the likelihood of earthquakes in the region, stating that they might possibly occur once every

42. *Robitnycha hazeta*, November 23, 1988.

10,000 years. Hlazovyi angrily attacked this line of thought with examples drawn from the study by the Velikhov Commission. He pointed out that there was evidence of a massive earthquake in the area only thirty miles south of the nuclear plant site in the year 63 BC. Another major earthquake occurred in the third century, destroying the ancient city of Chersonesus, originally believed to have been sacked by nomadic tribes. A large earthquake also occurred in 1751. At that time, he stated, the Turks who ruled this region built fortresses specially constructed to withstand earth tremors. A further ten earthquakes had occurred since 1790, including one in 1927 with a force of eight balls on the Soviet scale.

On September 26, 1988, Radio Moscow announced the results of the Commission's review, noting that the original project had anticipated seismic activity of six balls' intensity in this region, but the actual figure was closer to ten balls. After "comprehensive research" into the construction by the Commission, it had been concluded not merely that the project was unfeasible, but that it would be a "serious crime" to complete it. In an interview that followed this announcement, Vladimir Pivovarov, head of the Department of Solid State Physics at Simferopol State University, stated that in the event of an accident at the station, the whole of the Crimea and the Azov and Black Seas would be covered with radioactive dust. He added that the decision to construct the plant in the area had been made in the false security of the feeling that "nothing could go wrong," and he insisted that those responsible for making this decision, which had resulted in substantial central investment, should be punished.

Likewise, the accusation of negligence against those specialists who conducted seismic tests in the area in the 1970s was also raised by members of the twenty-one member working group appointed by the Velikhov Commission. The Commission also discovered that even had local officials wished to measure seismic activity around the station, there were not enough instruments available for them to do so. The working group was, however, not unanimous in its recommendation to abandon the Crimean project. It appears that fifteen condemned the station because of the high seismic levels in the region, two had some reservations, and four, who worked for the Ministry of Atomic Energy itself, wanted to complete the construction. It would seem therefore that ministerial interests predominated among the four dissenters. Hlazovyi pointed out that earthquakes were only one problem in the Kerch region; the area had also experienced volcanic activity. In 1930 and 1974, there were huge eruptions of the Dzharzhavsky volcano, on the edge of the peninsula. In September, 1988, there was also a volcanic eruption in the Azov Sea, only forty-two miles east of the plant site. As a further argument against the troubled edifice, he cited recent flooding at the site and negligent construction of the cooling reservoirs.

Following the investigation by the Government Commission, work on the Crimean plant was halted, pending further investigation by the Soviet government. One recommendation—offered by Hlazovyi, though he was not the first to do so—was to transform the current station into a training center for nuclear plant operatives, which would mean that the building was not a total loss to the economy. There then followed a period of almost total inactivity from the central authorities that infuriated Ukrainian environmentalists and academicians alike. By taking no action on the important study by the Velikhov Commission, the government revealed the divergence of opinion on the topic. At the very least, it can be said that there was a powerful lobby in Moscow that was extremely reluctant to abandon the Crimean plant. In contrast to the backdown on some other projects, the authorities baulked at dropping a station that was already so far advanced that the first reactor was practically ready to be commissioned. Curiously, the two most significant protests against this reluctance to abandon the project came from the scientific level.

The first came in the shape of an article in the newspaper *Robitnycha hazeta* by Ivan Lukinov, a vice-president of the Ukrainian Academy of Sciences, which was written on behalf of the Academy in response to the article by Hlazovyi. Lukinov maintained that Hlazovyi had "objectively and accurately" portrayed the essence of the problem. Citing the findings of the Commission, Lukinov stated that although the Academy did not possess the authority to demand that work on the station be halted, it believed that the facts uncovered during the investigation were sufficient to warrant the unanimous conclusion that construction should be halted, and the already completed buildings should be used for generating power using a different fuel.[43] Only three days later, a second and lengthier article was published by ten scientists in the central party organ, *Pravda*.[44]

This second article noted that the construction of the Crimean plant was begun in 1980 on the Kerch Peninsula, one of the best potential health resorts the Crimea has to offer, and everything went according to plan until the accident at Chernobyl. Thereafter, numerous letters and telegrams were sent to the organizers of the Nineteenth Party Conference (June 1988) expressing fears about the ecological safety of the site—specifically, the dangers of earthquake activity. Like Hlazovyi, the ten scientists pointed out that the original forecast that any possible earthquake would not exceed an intensity of seven balls on the twelve-point Soviet scale had proved to be an underestimate by at least two balls. The findings of the government commission, they stated, should be discussed and approved by a meeting of the Interdepartmental Council on Seismology and Seismic Construction with the Presidium of the USSR Academy of Sciences.

The Commission had argued, it transpired, that although the probability of an earthquake's registering nine points on the Soviet scale in the next decade was not high, even the remotest chance of such an occurrence could not be ignored in the case of so dangerous a structure as a nuclear plant. The Interdepartmental Council had met on September 20, 1988, and approved the findings of the working group by sixteen votes to four. In its final recommendations of November 28, the government Commission reiterated the danger from earthquakes, a factor that acquired even greater relevance as a result of the earthquake in Armenia one week later and the desire of experts to shut down the nuclear plant in Armenia ahead of the scheduled 1991 date.[45] The

43. *Robitnycha hazeta*, January 8, 1989.
44. *Pravda*, January 11, 1989. The article carried the names of an academician, a corresponding member of the USSR Academy of Sciences, two doctors and one candidate of geographical sciences, a doctor of economics, three doctors of geological-mineralogical sciences, and a doctor of physico-mathematics.
45. Velikhov himself cited statements by experts that the Armenian nuclear plant should be shut down at once unless immediate steps were taken to protect the station against future earthquakes. *TASS*, December 29, 1988. See also Elizabeth Fuller, "Date for Closure of Armenian Nuclear Power Station Brought Forward," *Report on the USSR*, No. 1, 1989, pp. 21–22.

Commission also noted the danger of volcanic activity around the site, and concluded that this was a further inducement to terminate construction.

The article by the scientists also cited a letter sent to Fedir Morhun, then Chairman of the USSR State Committee for the Protection of Nature, by the USSR Academy of Sciences Scientific Council on Problems of the Biosphere in June, 1988. The letter stated that, as early as 1987, the academy had been studying the possibility of banning the construction of new industrial enterprises and the expansion of existing ones in the Crimea. A decision had been taken on this matter in September, 1988. The goal was to keep a prime vacation area for Soviet citizens free of industrial enterprises with a high power requirement. The scientists thus believed that a strong case for halting the construction of the Crimean nuclear plant had been made to the USSR Council of Ministers, which had to make a final decision.

The problem, in their view, was the prevailing lack of activity at the highest level. The scientists had decided to publicize the findings in *Pravda* because in the past the opinions of experts had been ignored or distorted by various departments. An earlier reference in the text made it plain that the scientists were alluding to opposition from the Ministry of Nuclear Energy (founded in July 1986), some members of which were included in the Commission. They also pointed out that excessive enthusiasm for one form of power generation—first hydroelectric plants, subsequently nuclear plants—had led to major ecological and socioeconomic mistakes, delaying the development of energy-saving technologies. They felt key decisions were being taken by the USSR Ministry of Power and Electrification and Ministry of Nuclear Energy, often without up-to-date and objective information. What was needed, in their view, was a system of independent investigations, preferably by scientists from the USSR Academy of Sciences, at an early stage in the planning of major power projects. In short, the monopoly on decisionmaking should be taken away from the two ministries, which could not be trusted to make an impartial analysis of the siting of nuclear power plants. The article advocated close supervision of those involved in the planning and construction of nuclear plants. The scientists asserted that their views were in line with the general public, with perestroika, and with the precepts of Vladimir Vernadsky, one of the founders of the Ukrainian Academy of Sciences, thus representing the prevailing trend in the Soviet Union at this time against "bureaucratic cliques."

Again, a significant warning was followed by a period of seeming inertia. Yet such lack of action at the central level not only spurred the progress of economic sovereignty in Ukraine, but it directly encouraged the environmental movement. Along with the Chyhyryn station on the Dnipro River, the issue of the Crimean plant was to be raised at every major demonstration and protest in Ukraine. The nuclear authorities, while slow to react, did offer a response in the summer of 1989, when the Minister of Atomic Energy, Nikolai Lukonin, commented that the station would not be started up until it had been inspected by a team of international experts.[46] This somewhat defiant remark, implying that the station would eventually come into service, was soon negated by the removal of Lukonin from his post. The number of Crimean petitioners against the plant had by this time risen to 350,000, and the protestors had the backing of the Crimean party and government leaders.

In mid-September 1989, the First Secretary of the Crimean Oblast party committee, Andrii Hyrenko, wrote a letter to Deputy Chairman of the USSR Council of Ministers, Lev Ryabev—who had inspected the station—bemoaning the fact that construction work was continuing and noting that further building had been scheduled for the remainder of 1989.[47] Such procrastination had occurred in spite of public opposition, and as a result the building zone was said to be experiencing "unprecedented tension." People were incensed at the failure of the Crimean government to solicit prompt action from its Soviet counterpart. At a session of the oblast government, speakers criticized what they perceived as indecisiveness on the part of their leaders, and virtually every speech demanded a halt to construction work at the station. While the meeting was taking place, members of an ecological association, "Ekologiya i mir," gathered outside the meeting hall to demand an end to investment in and building of nuclear reactors in the Crimea.[48] Furthermore, it was reported that

46. *Radyans'ka Ukraina*, August 9, 1989.
47. *Pravda Ukrainy*, September 15, 1989.
48. Radio Moscow, September 27, 1989.

there was a serious threat of an all-Crimean strike in protest against the continuing construction of the plant. There was continuous picketing in the building workers' settlement, and the protestors wanted the site transformed into a training center for plant operatives or some other "ecologically harmless" facility.[49] The authorities finally relented under this unprecedented pressure, and the Crimean project was formally abandoned in late October 1989, to be reprofiled into a training center for nuclear plant operatives.[50]

SOUTH UKRAINE AND CHYHYRYN

A second area of major focus among Ukrainian nuclear projects has been the South Ukraine project, located in Mykolaiv Oblast, and being built by the USSR and East European countries, particularly Romania, which had had a significant investment into it. Originally scheduled to have four 1,000 megawatt VVER reactors, the ultimate capacity was extended to 8,000 megawatts and in addition a giant hydroelectric power complex, including three hydrostations and a hydro-accumulation project, was to be constructed on the same site. Environmentalists were thus particularly concerned about the impact of such a power construction on the rivers being harnessed for these projects, particularly the South Buh, one of the few rivers in the Ukrainian republic that were not totally contaminated by the mid-1980s. Since the background to the project has been elaborated elsewhere,[51] suffice it here to analyze the discussion around the future of the station in the Gorbachev period.

The debate took on a lively form in the Ukrainian press in the fall of 1988. It was sparked by an article written by V. Bilodid, an engineer at the South Ukraine nuclear plant, published in the

49. Radio Kiev, September 29, 1989.
50. *Izvestiya*, October 26, 1989.
51. See David R. Marples, *Chernobyl and Nuclear Power in the USSR* (London: Macmillan), 1986, pp. 80–83. Incidentally, Romania's only station under construction at Cernavoda, based on CANDU reactors supplied by Atomic Energy of Canada Limited, had been revealed to possess serious flaws in equipment and design. There were also unconfirmed rumors of the use of "slave labor" to complete the building under the regime of the late Nicolae Ceaucescu.

newspaper *Robitnycha hazeta*.[52] Bilodid was particularly worried that completion of the second stage of the nuclear plant (by which he signified reactors three and four) would cause irreparable damage to the South Buh river and its animal life. He also referred to the dangers of overheating the Konstantinovka and Tashlytske reservoirs, which were to be used for cooling purposes at the plants. Similarly, on November 11, 1988, another letter was published in the same source signed by L. Sharaev, the First Secretary of the Mykolaiv Oblast party committee. This stated that there was widespread concern about the building of reactors three and four, and outright opposition to the third stage—the construction of fifth and sixth reactors at the station.

Bilodid's article was criticized in a letter to the newspaper from the planners, who included V. Osadchuk, the Director of the Ukrainian branch of the All-Union Institute for the Design of Hydroelectric Power Stations (Gidroproekt), which was subordinate to the USSR Ministry of Power and Electrification, and the chief engineer of the project, L. Levytsky. Although Osadchuk and his colleagues had apparently sent the newspaper a twenty-two page response, "insisting" on its publication, the editorial board of *Robitnycha hazeta* printed the letter in a very abbreviated form. The newspaper offered the justifiable plea of shortage of space for abbreviating it, but then proceeded to devote considerably more space to its response than it had allotted to the letter.[53]

The planners pointed out that Mykolaiv Oblast was suffering from acute water shortages (an odd opening, surely, for a defense of a "water-hungry" nuclear power station!) and fuel-energy imbalances and that the power stations had been planned as early as the 1960s. They stated that various sites had been considered for the South Ukraine complex, including four for the nuclear power plant and ten for the hydro project. By 1970, a proposition had been put forward to amalgamate the power stations into one unit. Refuting Bilodid's remarks to the contrary, the planners maintained that the runoff waters from the nuclear plant were being localized in a special circulation system that was not connected to the reservoirs or the South Buh.

---

52. *Robitnycha hazeta*, October 14, 1988.
53. *Robitnycha hazeta*, December 18, 1988.

Bilodid had noted that there were technical illegalities in the organization of the South Ukraine plant's cooling system. According to Osadchuk and his colleagues, however, the cooling system at the plant did not contravene the new rules on the siting of nuclear plants approved by the USSR Council of Ministers in October 1987, since, they stated, the rules permitted a direct-flow system of cooling (as opposed to water that is recirculated). They added that the original plans had been approved by a commission from the Ukrainian Academy of Sciences in 1975 and that the second stage of the nuclear plant had been sanctioned in 1984–85, again after careful investigation. Having noted the economic benefits of the giant complex to the region, the planners then denied that the ambitious scheme would adversely affect plant and animal life, disputing the contention that the rivers flowing into the reservoirs contained rare plant life that would need transplanting. Finally, they attacked Bilodid's assertion that the plans for the complex had not been devised with sufficient caution, pointing out that they had been elaborated over the course of a decade (1975–85).

The editorial board of the newspaper itself offered a reply to the letter in a surprisingly aggressive approach that suggested that pro-nuclear power lobbies might already be in the minority in Ukraine. It was revealed that the oblast officials in Mykolaiv had long been trying to alert all-Union and Ukrainian government officials to the situation. The Mykolaiv government was quoted as expressing its "uneasiness over the numerous deviations from the plan in the construction of the first and second stages of the South Ukraine nuclear power plant." It was said to have appealed repeatedly to the USSR and Ukrainian SSR Councils of Ministers, the USSR and Ukrainian State Nature Committees, and the Ukrainian Academy of Sciences, but without avail. The oblast party and government were also reported to have sent another letter to the USSR Council of Ministers in November 1988, in which they demanded that an expert commission reexamine the project; that a decision be taken regarding the possibility of drawing up a new plan to look into alternative energy sources in southern Ukraine; and that all hydrotechnical construction on the South Buh River cease in the interim. Apparently, the oblast Prosecutor's Office also stepped in, noting that plans for the third stage of the nuclear power plant did not take into consideration the need to protect local supplies of fish.

According to the editorial response, the conclusion of many specialists was that extending the nuclear plant to a second, third, and fourth stage would cause serious damage to fish supplies and reduce the flow water of the South Buh to little more than one-third of its present level. This would lead, it was said, to a reduction of "natural spawning" by 80 percent and would bring the salinization of the water to a critical level. The newspaper cited the decision to abandon construction of the Danube-Dnipro Canal and suggested that the planners of the South Ukraine energy complex were simply presenting the same arguments that had been used by supporters of the now obsolete canal. In conclusion, numerous examples were provided of what the newspaper perceived as legitimate opposition to the energy complex, including a "concerned protocol" from a party meeting at Reactor Section No. 1 at the nuclear plant itself. Surely, the editorial board argued, planners should concern themselves less with the scarcity of electricity than with the need to economize on water use and conserve energy.

Subsequently, future stages of the project were postponed. The above debate was symptomatic of others taking place in Ukraine around nuclear projects: at Rivne and Khmelnytsky, which had been inspected by teams of international experts, but especially at Chyhyryn. The Chyhyryn station was abandoned permanently on May 19, 1989, when the USSR Council of Ministers announced that it was to be converted into an enterprise that was "ecologically clean" and excluded from the Soviet nuclear energy program.[54] The announcement followed an intensive public campaign against the station, which became the third energy-related enterprise to have been abandoned at the same site. The main arguments advanced against the station's construction were that it would be ecologically dangerous, as the station was to have used cooling water filtered from the Dnipro River—already drawn on to capacity; that building work had begun without the necessary inspection and without consulting the local residents; and that the location of the plant was a well known historical landmark in Ukraine, the headquarters of the former Hetman state.

54. *Radyans'ka Ukraina*, May 24, 1989.

RESPONSE TO ECOLOGICAL CONCERNS OVER NUCLEAR POWER

The problems of the Ukrainian nuclear power program were made apparent by other events in 1988–89 in particular. In assessing them, it is worthwhile to recall the basic points made in the Ukrainian appeal to the Nineteenth Party Conference in June 1988. It stated that there were strong objections within the republic to the construction of the Rivne nuclear power plant on a "karst" (permeable limestone) foundation; to the building of the Crimean station in a seismic zone; to the extension of the Khmelnytsky, Rivne and South Ukraine stations, primarily because of the insufficiency of water supplies; to the low quality of plant construction and the failure to protect the environment; and to the disregard of the psychological effects of the Chernobyl accident. A moratorium of ten to fifteen years on the construction of nuclear plants in the republic was proposed.[55] The ecological association "Zelenyi svit" (Green World, see below) published a six-point appeal in the newspaper *Vechirnii Kyiv* that made reference to three stations—Crimea, Chyhyryn and South Ukraine. In each case, an immediate cessation of construction work was demanded, pending judgment by a body of experts.[56]

The authorities, unaccustomed to public opposition to the nuclear power program, responded by asking the IAEA to inspect problematic plants. Thus in January 1989, IAEA representatives from eleven countries, including the United States and Canada, arrived in the Soviet Union to inspect the Rivne station. Despite some minor criticisms however, such as the remarks that equipment maintenance needed to be improved and that full-scale simulators should be used for operator training, the verdict was positive. It was pointed out that, while the construction of Rivne on karst had necessitated the expenditure of additional funds, there were nuclear plants in Brazil and Japan that were also built on karst. Overall it was considered that the solution to the public outcry about the safety of Rivne station lay in better communication with the people who must live in the vicinity. As Ferdinand Frantzen, one of the visiting scientists, commented, "when people are well informed, there are fewer anxieties."[57]

55. *Literaturna Ukraina*, June 23, 1988.
56. *Vechirnii Kyiv*, December 26, 1988.
57. *Pravda Ukrainy*, January 14, 1989. An international seminar on the situation

In May 1989, V. Fokin, chairman of the Ukrainian State Planning Committee and a deputy chairman of the Ukrainian Council of Ministers, published a lengthy article on nuclear power in Ukraine.[58] In summing up the related questions of protection of the environment and the development of nuclear energy, Fokin spoke of the decisions to "preserve" planned nuclear power and heating stations at Odessa and Kharkiv (a third such station in Kiev has been converted to a thermal power station); to stop the construction of the fifth and sixth reactors at Chernobyl and the building of units five and six at the Khmelnytsky and Rivne stations; and to halt the constructions at Chyhyryn and the Crimea. A slowdown had also occurred for the first time at Zaporizhzhya as a result of a safety inspection of the fifth reactor. It seemed as though in the aftermath of Chernobyl, the entire Ukrainian nuclear power program had ground to a halt, though there was never any official indication that the various problems were other than temporary setbacks, caused for the most part by emotional and overexcited public reactions.

Before the decision was made to stop the Chyhyryn construction, the most prominent feature of the Soviet nuclear power industry was its seeming ability to ride roughshod over public sentiment in Ukraine. Certainly mass petitions were ignored and the only major change in policy was a new Soviet adherence to IAEA recommendations not to build reactors close to city centers. The Soviet nuclear power authorities were, however, anxious to be perceived as responding to dangers. Thus in the spring of 1989, they agreed to adopt a revised "General Safety Rule," which corresponded to IAEA standards. Also, as part of an agreement first signed in April 1988, Soviet nuclear safety inspectors decided to visit the Catawba nuclear plant in South Carolina, USA, while an American team was in return invited to spend about seven weeks examining the Zaporizhzhya station.[59]

Both Soviet and international nuclear organizations have been created that emphasize attention to ecological concerns and to the safe operation of nuclear power plants. An International En-

at the Rivne nuclear power plant was held in Kiev in February 1989. See *TASS*, February 28, 1989.

58. *Robitnycha hazeta*, May 20, 1989.

59. *Associated Press*, May 19, 1989.

ergy Club, a nongovernmental organization that was to study the possibility of a unified energy system, was founded in Moscow. Formed under the aegis of UNESCO and consisting of nongovernmental associations of energy experts, ecologists and antinuclear war activists, it began to look into ecological problems, prognoses for energy development, and new techniques for conserving energy. It held its inaugural meeting at the International Trade Center in Moscow from March 1 to March 6, 1989.[60] In February 1989, nongovernmental technical specialists were invited to join a newly created Nuclear Society of the Soviet Union. Then in March, Soviet Premier Nikolai Ryzhkov addressed a conference of the World Association of Organizations Operating Nuclear Power Stations, a body whose establishment the Soviet government reportedly considered a significant step toward uniting efforts to create an international regime for the safe development of nuclear energy. As the title of the organization suggests, it was strongly in favor of nuclear power, which it maintains is ecologically the cleanest form of energy production.[61]

To the public, however, such amendments were inadequate. The sentiment against nuclear power was so widespread that reasoned arguments were immaterial. Indeed, virtually every proposed scheme was countered with the simple response: "Chernobyl." The rising costs of the Chernobyl disaster had already rendered the Soviet nuclear power program economically unviable; they had negated statements that nuclear power stations enabled energy saving. News and rumors about past accidents had also become rife. And, as we have seen, some of these protests appeared to be justified, especially in the Crimea. Glasnost and the development of public opinion in Ukraine heightened this process. While Ukraine was relatively slow to introduce the processes of glasnost, it was further advanced, for example, than Belorussia, where the dangers from Chernobyl radioactivity were even greater, but public ignorance about the fact was preserved for much longer. Ukrainians were becoming militant on the question of ecology and they were not prepared to accept safety precautions made in conjunction with the IAEA (an organization that was ipso facto in favor of nuclear energy) or to sit back and

60. *TASS*, January 11, 1989, and March 1, 1989.
61. *Izvestiya*, February 9, 1989; *TASS*, May 16, 1989.

watch what were seen as timid and cautious efforts of the USSR and Ukrainian State Nature Committees to ensure that ecological concerns were included in the planning and operation of any industrial enterprise. The end result, therefore, of Chernobyl, Narodychi, irrigation and nuclear power schemes was a Green World association and a Green Party, the first an informal ecological group; the second a party that could participate in the complex political situation that was emerging in Ukraine.

5 | ECOLOGY
The Green World

THE FORMATION OF THE Green World ecological association (and subsequently also a Green Party) was a response not
merely to Chernobyl, irrigation and nuclear power schemes, but
to a whole host of ecological problems facing Ukraine that had
received little attention for years. In fact, the formation of the association was in part a response to the lack of official action on
such subjects. Glasnost and the freedom of Ukrainians to travel
abroad raised this awareness of nature and the environment,
while Chernobyl created in the public a sense of responsibility, a
belief that it was possible to control one's own destiny and that
otherwise a catastrophe was imminent. The impetus came largely
from Ukrainian writers and intellectuals, both inside and outside
the Communist Party. The Ukrainian Union of Writers was perhaps in the late 1980s the most radical sphere of society. Even in
the Stalinist period, writers had been freer to express their views
than others. In Ukraine, where a hardline party organization held
firm sway over the population even during the first Gorbachev
years, it was the writers who addressed many of the prominent
problems facing Ukrainians.

THE 1988 ECOLOGICAL DEMONSTRATION

One can officially date the beginning of the Green World association to December 1987, but the start of its impact upon society
could well have been an officially organized mass demonstration
that took place in the city of Kiev on November 13, 1988, in

which an estimated 10,000 people (some sources put the attendance much higher) took part. The demonstration was permitted to take place because it was to have focused exclusively on the ecology. But it also represented one of the first occasions when the transition from ecological to political questions presented itself. Thus speakers attacked the Kiev party leadership as a "Brezhnevite" party apparatus, particularly Shcherbytsky, the First Party Secretary. Speakers interviewed about the event stated that they regarded the demonstration as the first significant step toward strengthening the Popular Movement to Support Perestroika (Rukh), which was officially formed in 1989. One reason why this should be the case in Kiev was that similar attempts to form the Rukh in Lviv had been harshly repressed.

The November 13 demonstration was organized by four groups: the Ukrainian cultural group called *Spadshchyna* (Heritage); *Hromada*, a student group from the University of Kiev; the Green World; and an informal ecological association called *Noosfera*. Speakers at the meeting included well known literary figures such as Yurii Shcherbak, Volodymyr Yavorivsky, and Dmytro Pavlychko; members of the Ukrainian Helsinki Union, such as Oles Shevchenko and Ivan Makar; representatives of the "green" movements and the Democratic Fronts from Latvia and Lithuania; experienced academicians, such as the Russian, Fadei Shipunov; and activists from other parts of Ukraine, such as the poet, Rostyslav Bratun, at that time a representative of the "Lions Society" from Lviv.[1]

The Ukrainian press provided partial and subjective coverage of the demonstration. Reports appeared in the newspaper *Robitnycha hazeta*, which had a growing reputation for its concentration on environmental issues; and in two daily newspapers that have remained unavailable in the West, *Vechirnii Kyiv* (published also in a larger, Russian edition) and *Prapor komunizma*.[2] In *Vechirnii Kyiv*, which published the most detailed account, only

1. This account is based on a tape of the demonstration provided to the author by Chrystia Freeland, a Canadian student, who was present in Kiev; and upon eyewitness accounts given to the author in Kiev on the following day.
2. *Vechirnii Kyiv*, November 15, 1988; *Prapor komunizma*, November 15, 1988; *Robitnycha hazeta*, November 16, 1988. Incidentally, although *Vechirnii Kyiv* is known to be much more radical in its outlook than *Prapor komunizma*, both are housed in the same building in the outer reaches of Kiev and the staff of both papers intermingle freely.

selected speeches were presented, and the more controversial statements at the meeting were omitted. The focus was largely on speeches by Shcherbak and Shipunov—arguably two of the most important delivered, however—and the Baltic and more political speeches were omitted. Proceedings began with a speech by Pavlychko, who demanded a stoppage of construction work on the Chyhyryn nuclear power plant.

Shcherbak began with the subject of Chernobyl, announcing that he had just returned from visiting Prypyat and radiation victims in their hospital beds elsewhere. This radioactive fallout, which constituted, he stated, one of the biggest ecological catastrophes in the history of civilization, had spelled an end to myths and illusions about the reliability and infallibility of science and technology. At the same time, he berated official secrecy about the catastrophe, noting that the public was forced to rely on information in foreign medical journals about the dangers of radioactive hotspots. Denouncing other nuclear projects in Ukraine, he demanded an authoritative public commission made up of all sectors of the republic to study the effects of Chernobyl. He also sought public control over the work of the Ministries of Health and Atomic Energy; the commemoration of April 26 as the Day of Remembrance of Chernobyl, and the erection of a memorial to its victims in Kiev. Perhaps most important, he demanded an international forum on Chernobyl and a referendum throughout Ukraine on nuclear power. Thus the significance of Chernobyl on Shcherbak, at least, was manifest.

He was followed to the podium by F. Ya. Shipunov, head of a laboratory at the Scientific Council on Questions of the Biosphere with the USSR Academy of Sciences in Moscow. Shipunov declared that Ukraine was approaching an ecological catastrophe, which was coming closer with every passing day. The Ukrainian people, in effect, he stated, were living inside a nuclear reactor. He attacked Soviet nuclear energy stations in particular for the fact they released into the atmosphere some 350 percent more tritium than other plants, a biologically active substance. "Anti-ecological systems" had been created on the Volga and Dnipro rivers. The various energy projects on the Dnipro, he pointed out, may be recouping profits of some 100 million rubles annually, but the amount of damage they were causing to the environment amounted to over 25 billion rubles annually. The atmosphere in Kiev and other major cities was "dead." In fact space-craft orbiting Kiev and other cities over the previous three years had revealed a dangerous reduction in the concentration of the ozone layer.

Shipunov next turned to agricultural reform, which he said could begin with the liquidation of more than 50 percent of the ministries and associations. A land reform was required in both Ukraine and Russia, which recognized that the basic agricultural unit in the country should be the family cooperative organization. Finally at the oblast and raion level, he stated, ecological committees should be created, which would control economic schemes that threatened the ecological welfare of the public. Among the other significant speeches that focused on issues not related to Chernobyl was one by Rostyslav Bratun, present in his capacity as the Chairman of the Green World branch in Lviv Oblast. He attacked the construction of the Rivne nuclear power plant on dangerous foundations (see the previous chapter), and the construction of giant chemical plants in Western Ukraine, in cities such as Drohobych, Ivano-Frankivsk and Kalush. When planning such constructions, he maintained, the public must be allowed a voice, and a green Carpathians must be permitted to survive.

Following such angry protests, the audience was evidently in no mood to tolerate statements to the contrary. Yet protocol demanded that a hearing be given to a representative of the State Committee for the Protection of Nature, who acknowledged that the environmental situation in Ukraine had become complex, mainly because of the economic emphasis on machine building and the chemical industry there. But, he stated, nature protection organs were being formed in all oblasts and raions of the republic. The amount of pollutants in the water flow from factories into rivers, he declared, had been reduced by almost two times over the period 1986–87, and over one billion rubles were being expended annually in Ukraine for the protection and rational use of water resources in the polluted basins of the Black and Azov Seas. Even in cities, in which automobile transport was the major pollutant, he maintained, the ecological situation was stabilizing (there was reportedly a gasp of disbelief among the audience at this point).

This unpopular comment was followed by a bitter denunciation of the situation in Cherkasy by Yurii Vysotsky from the "Ecology" society. The latter had circulated and published in several major newspapers a letter on the "ecological degradation of Cherkasy," focusing on the building of the nuclear power plant there (Chyhyryn) and other industrial problems. The letter stated that citizens were tired of hearing about glasnost or about politics and wished instead to hear about ecology. A more factual presentation was offered by S. Dorohuntsov, the Chairman of the Council for the Study of the Productive Forces of Ukraine, who has had the task of making prognoses about the development and locations of industries to the year 2009 and beyond. The problem of electricity production in Ukraine was the most acute, he declared and how much was necessary for Ukraine. Per head of population the amount produced in the republic approached the level of a major Western industrial power such as West Germany. Twelve nuclear reactors were functioning, but a further twenty were under construction or planned. Chyhyryn would supposedly not be built, but no one seemed to have advised the construction crew, since residences and other amenities were being erected.

A second problem, in Dorohuntsov's view was grandiose construction projects. The Dnipro-Buh estuary was a case in point. The Ukrainian Academy of Sciences together with the Mykolaiv government had opposed the project, but an estimated seventy million rubles had been expended (presumably by the Minister for Land Improvement and Water Economy) before the project had been officially approved. He anticipated severe and irrevocable damage to the Dnipro River if the project went ahead. Meanwhile the Ukrainian government in his view was bent on a major expansion of the chemical industry, anticipated a 130 percent expansion over fifteen years, following an expansion rate of 240 percent over the period since 1965. Other speeches continued in this vein, returning the emphasis to nuclear power. However, what had been expected to be a four-hour meeting was to be cut short abruptly after two and a half hours when Ivan Makar of the Ukrainian Helsinki Union stepped forward to speak (he had been released from prison only a few days earlier for participating in an August 4, 1988 demonstration in Lviv).

When Makar began to talk, security officials, who were present in large numbers, switched off the microphones, forcing him to shout his speech.[3]

Without a microphone, Makar argued that ecological concerns should not be separated from socio-political matters. He maintained that it was important to emphasize the seriousness of the "draconian" laws that the Supreme Soviet was about to vote on (the amendments to the Soviet Constitution that would, it was felt, in effect nullify the right of republics to secede from the Union). He went on to say that the coercion of Ukrainian citizens was continuing and that "our republic will not be granted any sovereignty." He suggested that Ukrainians should form "a truly national front" and closely align their actions with those of the peoples of the Baltic republics. In doing so, he continued, Ukrainians could resolve their cultural, language and economic problems. If they did not work in close cooperation with the Balts, holding even thousands of demonstrations would have little or no effect on party chief Shcherbytsky or on the Minister of Health of Ukraine, Anatolii Romanenko.

Such speeches clearly aroused the ire of officials who felt that they went beyond supposedly apolitical ecological concerns. As will be shown, however, the dividing line between the latter and politics was often a very thin one, a tightrope that Green World leader, Yurii Shcherbak, would attempt to tread valiantly. In the account in *Robitnycha hazeta,* the authors described some speakers who unsuccessfully tried to thrust "all their cheeses into one dumpling." They went on to say that in addition to the well-meant and sincere speeches given at the demonstration, there were other speeches that amounted to outright demagogy. On the other hand, the newspaper's report was also very critical of those people at the demonstration who, inexperienced in "cultural democracy," had attempted to prevent speakers from voicing their opinions.[4] *Robitnycha hazeta* tried to take a middle line between what it perceived as the two extremes represented at the Kiev meeting: manifestations of anti-party nationalism (the at-

---

3. The event was captured in Shklyarevsky's 1989 film, "Mi-kro-fon!," the title of which was derived from the chant of the audience after the microphone was disconnected.
4. *Robitnycha hazeta,* November 16, 1988.

tempt to form the Rukh, for example) and Stalinist-style repression.

At the end of the meeting, the organizers presented a series of resolutions that were neither mentioned nor published in the two Kiev-based newspapers, but were cited in *Robitnycha hazeta*. The resolutions formed the basis of a petition that was to be circulated in the republic and eventually sent to the Ukrainian Supreme Soviet. They covered both ecological and political issues. The ecological situation in the world, and in Ukraine in particular, was described as serious and in need of immediate attention. It was pointed out, moreover, that the party apparatus in the Ukrainian SSR did not represent the interests of the Ukrainian people and had thoughtlessly exploited natural resources. The resolutions contended also that the whole truth about the Chernobyl disaster had still not been revealed and that the authorities were continuing to propagate the "myth" of the lack of alternatives to nuclear power. First and foremost, the demonstrators demanded that construction be stopped at the nuclear reactors at the Crimean and Chyhyryn stations and that no new nuclear reactors be built in Ukraine. It was stressed that the expansion of existing stations must be brought to a halt and that the three reactors now functioning at the Chernobyl plant should be switched off and the plant closed down completely.

The resolutions called for those reactors currently in operation in Ukraine to be examined by experts in order to ascertain whether their technology met international safety standards. Furthermore, other industries that were ecologically harmful and located in densely populated regions must be closed down and future construction curtailed. The resolutions focused on the South Ukraine energy complex. It was maintained that no further work should be undertaken on this complex until ecological studies had been conducted. On the general question of ecology, the resolutions asked for the "veil of secrecy" over ecological issues to be lifted. Only two days before the rally, it was reported that a chemical factory outside Kiev had emitted poisonous gas into the atmosphere. The only official account of this incident had been a newspaper report denying that there had been an accident.[5] A

---

5. Information of Oles' Shevchenko, at that time the Kiev leader of the Ukrainian Helsinki Union. In 1990 he was elected a deputy to the Ukrainian Supreme Soviet. Interview with Oles' Shevchenko, Kiev, November 15, 1988.

special ecological bulletin was to be established in Ukraine and there was a call for referenda to be held on matters pertaining to ecology.

The significance of the November 1988 demonstration was that for the first time, Ukrainians had united on the issue of ecology and expressed similar concerns, whether the problem was in Odessa, Lviv or Donetsk. Indeed, such agreement was unique. Politically Ukraine had become divided between an extremely recalcitrant and separatist West; a heavily russified East, the traditional stronghold of the party apparatus; and an equally disaffected but not so nationally conscious South. It had become evident that in making plans for the expansion of industry in Ukraine, the authorities had spared no zone, no city on considerations of nature protection or historical preservation. Moreover, it was also clear that in many cases, the local party and government associations concurred with the public's anxieties over the environment: at the least, they declared this to be the case. The result was a gathering of local forces in open opposition to centrally based ministries that were more concerned about fulfilling plans. Also, while some of the speeches at the demonstration represented an outpouring of emotion, many were carefully thought out and factual. There were so many individual ecological predicaments in the republic that a brief survey is necessary to elucidate the situation.

INDUSTRIAL POLLUTION

Industrial pollution had become the focus of attention in several parts of Ukraine. In areas as far apart as Rivne in Western Ukraine, Odessa in the south, Kiev in the center, and Dnipropetrovsk in the east, those concerned for the local environment had complained either about existing or planned factories and enterprises. The protesters' arguments were often similar, but they were pessimistic about their chances of success because of an antiquated process used to decide whether to build large industrial enterprises. Such a procedure permitted planners to ignore recommendations by independent expert commissions (as was the case with the Crimean nuclear plant for over a year) that had examined the feasibility of various projects.

In the summer of 1989, M.P. Skrypnyk, Chairman of the Ukrainian Committee for Hydrometeorology, published a concise survey of the ecological situation in Ukraine.[6] From an analysis of forty-two cities, he ascertained that in 1988, some eleven million tons of poisonous substances, including sulfurous oxides and ammonia, were released into the atmosphere from "stationary objects." The most serious offenders were located in the Donetsk, Zaporizhzhya, Dnipropetrovsk, Crimea and Kiev oblasts. Of the fifty most polluted cities in the Soviet Union, eight were in Ukraine: Donetsk, Dnipropetrovsk, Zaporizhzhya, Kiev, Kerch, Makiivka, Debaltseve and Dniprodzerzhinsk (though the omission of Mariupil from this list can only be regarded either as an oversight or truly astonishing). In these cities, automobile fumes accounted for most of the pollution. In Komunarsk, however, concentrations of ammonia in the atmosphere were more than twice the permitted level, while in Kiev and Mariupil they were more than five times the permitted level.

In early February 1989, the USSR Ministry of Ferrous Metallurgy was taken to task for its reluctance to deal with the huge Petrovsky metallurgical works in Dnipropetrovsk, which was polluting the surrounding environment. The factory (visited by Mikhail Gorbachev in 1985) was founded in the nineteenth century on the outskirts of Dnipropetrovsk, then called Ekaterinoslav. By 1989, however, it stood in the middle of a residential zone, and discharged dangerous amounts of toxic substances into the atmosphere over a radius of four kilometers around the factory. Three residential complexes were being subjected to concentrations of carbon and nitrogen oxides and to sulfurous anhydride that were two to three times the permissible norms.[7]

The most disturbing side effect of this contamination was the state of health of infants. Over the previous few years, the rate of mortality from congenital abnormalities in the area was said to have become exceptionally high. When compared with neighboring October Raion, the Lenin Raion, in which the factory is located, had a rate of respiratory infections in infants under one year that was up to eight times as high. In the outflow water from the Petrovsky works, the content of oil-based products was

6. *Radyans'ka Ukraina*, August 22, 1989.
7. *Stroitel'naya gazeta*, February 1, 1989.

ten times the maximum permissible and seven times that for iron-based products. Downstream from the works was the Shevchenko Park, said to be a favorite bathing area for local residents, but subsequently polluted. Consequently, radical reconstruction of the works was demanded. Bowing to pressure, the USSR Ministry of Ferrous Metallurgy came up with three possible variants for a remodeled enterprise. Of these, the third variant, which entailed restricting operations to a rolling section, while importing steel from the Krivyi Rih steel combine, was regarded as the most preferable. The ministry was portrayed as an unwilling agent of change, citing the argument that "the country needs metal." But was not the ministry asking too high a price, the correspondent for *Stroitelnaya gazeta* wanted to know?

A similar story came from neighboring Voroshilov Oblast (now called Luhanske), in the city of Komunarsk, which had a population of over 120,000. Here, the snow was said to be black by the time it reached the ground as a result of poisonous gases emitted by a metallurgical combine and a chemical plant.[8] Every year, two tons of dust from coal and metal fell from the atmosphere, making "mournful ecological arithmetic." One metallurgical combine was responsible for ejecting 145 harmful substances into the air. The author of an article in *Pravda Ukrainy* was incensed at how little was being done to alleviate this situation. An experimental gas filter, which took ten years to construct and cost about 500,000 rubles, had been forgotten. Nature conservation committees at the open-hearth sections of the combine were said to exist more on paper than in reality. At the chemical plant, a biochemical department created to clean up poisonous gases had been built without many of the necessary parts. The city council had created a program called "Ecology," but it had remained "in the dusty archives," according to the article.

Turning to future projects, one of the most serious conflicts emerged near Odessa, where a chemical combine was to be built on the bank of the Tilihulska estuary, near the settlement of Berezivske. The planners, it was reported, had called this enterprise a "factory," in order to conceal the immense size of the proposed construction. Among its tasks was to have been the

8. *Pravda Ukrainy,* January 25, 1989.

production of superphosphate. The Berezivske combine was to be located about thirty-seven kilometers from another chemical combine which, it was alleged, was already causing damage to the Black Sea and adjacent territories. The article, written by E. Taubman of the Odessa Technological Institute, pointed out the devastation that the project would bring to fertile agricultural lands and a recreation area, noting that such sites are already in very short supply in the southern part of the Soviet Union.[9]

Professor Taubman studied the feasibility of the combine together with a group of experts. He emerged from the exercise totally disillusioned with the project and its planning. The entire concept, he wrote, was based on experiments undertaken at small research institutes, the output of which was only one-tenth of what was proposed at Berezivske. There were nine factories producing superphosphate at present, he stated, all of which had poor economic returns and were ecologically harmful. Why not reconstruct these rather than build anew? Moreover, an analysis carried by Dorohuntsov had indicated that the enterprise at Berezivske was not even needed. Many questions raised by experts about the project had remained unanswered. For example, where would the 20,000 cubic meters of water "consumed" daily by the combine be drained off? What would be the makeup of raw materials discharged into the atmosphere? The documents supporting the construction of the Berezivke combine belonged in an issue of (the satirical magazine) *Krokodil*, declared Taubman. Thus the project had been approved on the basis of documents provided by the planners themselves rather than an independent analysis.

Of concern to Professor Taubman was the possibility that cadmium, a cancer-causing agent, would be one of the byproducts of the combine. How could one contemplate building an industrial enterprise, he asked, which might contaminate local fields with cancer-inducing substances? Yet such an outcome was not even anticipated in the documents for construction. Above all, Taubman complained about a planning system in which, in spite of the obvious defects of the plan, the unfortunate location of the building, and the overwhelmingly negative opinions of scientists and the public about the Berezivke combine, "construction will

9. *Robitnycha hazeta,* January 19, 1989.

soon be started." As a result of inherent shortcomings in the planning of large industrial enterprises in the Soviet Union, he asserted, poor quality constructions were frequently erected on sites that were ill suited to them. In the case of the Berezivke combine, the project was eventually abandoned, but it remains an instructive example of what was happening in Ukraine's chemical industry.

The city of Kiev also had not escaped ecological difficulties (in addition to the obvious Chernobyl-related problems), and particularly with the storage of highly toxic industrial waste, much of which was being dumped illegally on vacant plots of land. A decree on the building of a burial site was adopted in 1984, but had never been put into operation. The Kiev Oblast government conducted a survey on the question; the author of the report, Deputy Chairman V. Synko, concluded that, in order to avoid contamination of the environment around the city, the burial site should be within the thirty-kilometer zone around the damaged Chernobyl reactor, since this zone was already contaminated. The choice evidently brought little comfort to those worried about the state of the Dnipro and Prypyat rivers and of the Kiev Reservoir.[10] It should be added that neither at this time nor at present has the thirty-kilometer zone been entirely uninhabited.

Critics felt that the selected site—which was supposed to be ready by 1993—could contaminate other areas connected with the Dnipro River. In the view of the article's author, the best solution to such a project was to be found at the small Hungarian settlement of Azol, near Budapest. At a site there, a public commission had been created that had worked in close cooperation with the builders on the question of toxic waste. As a result, contaminants were said to have been kept safely and the public provided with a detailed account of each stage of the project. In Ukraine, however, such deference to the people's views had yet to take place.

In such situations, the public regularly sought redress by writing to local or Kiev newspapers. One such letter arrived at the editorial office of *Molod Ukrainy* in early 1989 (published on January 26), signed by 159 residents of the small village of Porubky (Rivne Oblast), who complained about the proposed

---

10. *Robitnycha hazeta*, January 28, 1989.

building of an ammonia plant one kilometer from the village. The writers maintained that the distance from the village was too small and that local fields would be polluted with nitrates and pesticides. The editors of the newspaper added a postscript that placed them squarely behind the 159 signatories of the letter. It was a crime to build such objects, they asserted, without proper ecological expertise and without lengthy discussions with those who were obliged to live in the vicinity. The builders were exhibiting a "cynical disregard" for the view of the Porubky residents. If those in favor of the project had some serious arguments to offer, then why had they not met the public to elucidate them?

The situation in Dniprodzerzhinsk was also giving cause for great concern. Industrial pollution was believed to be the cause of the alarmingly high mortality rate among newborn babies and of diseases, particularly of the skin, that had affected up to 25 percent of young children and 33 percent of schoolchildren in the city.[11] Earlier in the year 1989, at the behest of the newspaper *Radyanska Ukraina,* an independent association called "Ecological Initiatives" was founded in the city, made up predominantly of young people, including doctors, engineers and students. The association mounted a campaign to elect a deputy to the People's Congress but succeeded only after a strong opposition from local party authorities.

In Dniprodzerzhinsk and other cities, the key problems lie with metallurgical and chemical enterprises that have released more and more phenol, hydrogen sulfide and ammonia into the atmosphere. Pollution in Mariupil was said to be the worst because over the previous three years there had been no attempt to remedy the problem. There, the average concentration of nitric acid in the Kalmius and Kalchyk rivers was fourteen times the permissible maximum, that of oil products twenty-six times, and that of copper and sulfates eleven times. Huge steelworks such as "Azovstal" (Azov Steel) and "Zaporozhstal" (Zaporizhzhya Steel) had endlessly procrastinated over the introduction of measures to protect the environment that were now rescheduled for the post-1995 period.[12]

Much of the equipment in use at factories and steelworks had become outdated. For example, there were two blast furnaces at

11. *Robitnycha hazeta,* April 27, 1989.
12. *Robitnycha hazeta,* July 7, 1989.

"Azovstal" that had been in operation for forty-three years. A coke and chemical works in Dniprodzerzhinsk had been operating without a unit for biochemical purification of water. In addition, where purification units did exist, they were often switched off at night to preserve electricity, and thus contaminated products were released into the atmosphere or the water supply. Skrypnyk stated that after such emissions people in the nearby city had difficulty waking up the next morning (a common complaint also in 1990). There had already been a large number of accidental releases of poisonous substances, including sixty-eight in the first four months of 1989 alone by the Dniprodzerzhinsk production association "Azot."[13] Many observers perceived that the main culprits for such a situation were the Ministries of Ferrous Metallurgy and of the Chemical Industry, which did not take into account regional ecological questions when formulating economic plans.

Releases of hydrogen sulfide had also led to a particularly worrying situation in the Black Sea. A letter sent by a Kiev engineer to the newspaper *Pravda Ukrainy* in the summer of 1989 inquired whether there was any truth in reports that the Black Sea was about to explode because the solution of hydrogen sulfide in the water had reached saturation point.[14] If at some critical moment, he wrote, "some maniac were to strike a match," then an entire ecosystem would be destroyed and the country would be faced with another Chernobyl. Academician Valerii Belaev of the Ukrainian Maritime Hydrophysics Institute responded to the letter, assuring the writer that the Black Sea was far from saturation point, but he was less convincing in his descriptions of what might happen in the event of an earthquake. The idea of pumping out contaminated water from the Black Sea had been widely discussed but, according to Belaev, was impractical because it would require the construction of a giant industrial establishment with a huge infrastructure. Paradoxically, such a scheme would be prohibited in this region on ecological grounds.

According to journalist Volodymyr Kolinko, one of the authors of the film "Mi-kro-fon!," the Black Sea had reached a

13. *Ibid.*
14. *Pravda Ukrainy,* July 16, 1989.

critical stage by late 1989. He noted that 90 percent of the sea could already be declared "dead," a victim of hydrogen sulfide gas that had continued to rise from the depths of the sea and contaminate its upper layers. He pointed out that, at the end of the nineteenth century, this "dead zone" could be found at a depth of 200 meters; by the 1940s, it had risen to 125 meters; and in 1989, it was at the seventy-meter level. In his view, the Black Sea would be destroyed totally by the year 2040 if the present trend continued. What were the sources of this degradation of a major vacation area for Soviet citizens? Some of the hydrogen sulfide would have been produced "naturally" on the sea bed, but the situation had been made more complex by industrial developments in the Crimea, such as the Sivash aniline works, the Crimean titanium dioxide factory, the Perekop bromine factory, and the Crimean soda factory.[15] The Azov Sea was evidently in little better shape and one account emphasized the importance of ensuring that it did not share the fate of the Aral Sea.[16]

Dmytro Grodzinsky has conducted a revealing survey of the ecological situation in Ukraine from his perspective as a leading biologist. He has noted the severe anthropogenic load in industrial centers, the increasingly catastrophic state of landscapes, water, air, flora and fauna. Ukraine had been allocated only about 16 percent of nature preservation funds, but it accounted for 25 percent of pollution from waste and byproducts. One of the major difficulties has been that industry and agriculture had been developed extensively rather than intensively; literally the most easily available resources have been used for the economy. One might spread the area of a coalfield, for example, by extracting easily accessible coal over a wide region rather than develop one particular area. The amount of "hard" waste produced had reached 1.5 billion tons annually and has been scattered across some 230,000 hectares of fertile agricultural land.[17]

15. *Ukraine*, No. 11, 1989, p. 14.
16. *Robitnycha hazeta*, December 13, 1989. See also *Robitnycha hazeta*, March 30, 1990.
17. D.M. Grodzinsky, "Ecological Anxieties of the Ukraine," paper delivered at Rutgers University, Newark, New Jersey, February 17, 1990. Grodzinsky's remarks on the Dnipro River, following, can be supplemented by the conclusions reached by the Slavutych expedition, which demanded that no further industrial and energy projects be harnessed using the river as the basis. See *Radyans'ka Ukraina*, March 16, 1990.

A particularly acute problem had arisen with water, largely as a result of the "extensive" development of water projects. Thus the total volume of water consumption for industrial needs rose by 2.2 times over a twenty-five year period, but the volume of irretrievable water increased by 2.8 times for the same timespan. Water pollution had arisen in both major and minor river systems, but especially in the Dnipro, Dniester and South Buh rivers. Grodzinsky pointed out that in Ukraine, between 198 and 248 cubic meters of water were used to produce one ton of steel in the Donetsk region, whereas in West Germany, the total was only 180 cubic meters per ton. On the land, excessive plowing in Ukraine (57 percent of the territory of which is arable land) had caused soil erosion in 5 percent of the agricultural area, causing a decline in the crop yields of 15–40 percent. Plowing riverside regions had led directly to the destruction of 4,000 out of a total of 22,000 small rivers.

How had such a situation arisen in Ukraine? Grodzinsky has listed several principal causes of the "ecological crisis." First, the introduction of ecologically inept economic mechanisms and the exploitation of Ukraine's natural riches by monopolistic ministries from outside the republic. Second, the bureaucratic nature of society and centralization of major industries dominated by people who did not take ecological considerations into account. Third, poor technology in many branches of industry and agriculture. Fourth, the "dehumanization of society," and its development based on industrialization. Finally, he perceived an increasing division between the general cultural and moral level of the public and the growing "technologization" of life. In brief, then, Grodzinsky's view was that environmental problems had arisen largely from the nature of Soviet society, and made worse in Ukraine's case because of the republic's natural resources in both agriculture and industry.

One of the most tragic environmental accidents occurred in 1988 in the Bukovynian city of Chernivtsi, when schoolchildren began to go bald, apparently through exposure to a little known chemical. The first manifestation of this illness, which was accompanied by nervous disorders and pains in the limbs, occurred in August 1988, but little attention was paid to it at first. Eventually, a government commission was established to look into the situation under USSR Deputy Minister of Health, Aleksandr

Baranov. Altogether, about 150 children were hospitalized because of this "alopecia" between August and November 1988,
while the remaining children were largely evacuated from the city
so that the cause of the illness could be determined. The conclusion of the government commission initially was that the children
had been exposed to thallium, which may have come from factory byproduct releases or even from additives to gasoline tanks.[18]

The illness spread mass panic in the city. Eyewitnesses reported that many families simply fled, and touts offered train
tickets to Kiev at the station. By mid-November 1988, only 559
preschool children out of 15,000, and 5,000 schoolchildren out
of 33,000, remained in the city. The suffering children were hospitalized in Kiev, Moscow and Leningrad and evoked sympathy
from the pathetic photographs that appeared on the pages of
many major newspapers, forlorn, bald and helpless. At least two
riots occurred in the city because of public dissatisfaction with
the state of the government inquiry, and rumors were afoot not
only that the illness had spread beyond the borders of Chernivtsi
Oblast—into Moldavia, for example, but that adults were also
beginning to suffer from the same symptoms. Quite often, cases
arose of partial baldness, and the health authorities were very reluctant to suggest that this development was related to the alopecia that had affected the children.

After November 1988, the health authorities announced that
no more cases had been diagnosed. The city had been thoroughly
cleaned, with a curfew being placed on the movement of trucks
at night, and topsoil in city parks had been replaced. The children's hair, it was reported, had begun to grow back again. Yet
their symptoms, such as regular nightmares, often remained. The
numbers of those sick, moreover, continued to rise. From October 1, 1988 to October 1, 1989, 343 cases of alopecia were registered in the city of Chernivtsi, and 494 in Chernivtsi Oblast as
a whole (this latter figure would have included the 343 city
cases). Similar cases, it was stated, had occurred well outside the
confines of the region. International teams from bodies such as
the World Health Organization had been called in to examine

18. See, for example, Kathleen Mihalisko, "Disturbances and Unresolved Questions in Chernovtsy," *Radio Liberty Research Bulletin*, RL550/88, November
30, 1988. A good account of the problems in Chernivtsi also appeared in
Literaturna Ukraina, No. 46, 1988.

the children, but had been unable to offer much help. Major conflicts had begun to take place between informal groups and the party and government authorities because of the failure to pinpoint the source of the affliction. The only real consensus was that industrial pollution was the ultimate cause of the illness.[19]

But was thallium the cause of the chemical poisoning? One account noted that the USSR Ministry of Health latched onto this possibility "as if it were a lifebelt." Evgenii Chazov, the USSR Minister of Health suggested that the thallium may have been contained in acid rain, which had fallen on the city. However, water samples taken to several Moscow institutes, and examinations of the children affected, failed to uncover any traces of thallium. Experts on gasoline technology also largely discounted the notion of thallium being contained in additives to gasoline tanks. A massive rise in the concentration of boron in the city was detected by the All-Union Research Institute of the Ministry of Internal Affairs, and subsequently confirmed by the Institute of Nuclear Power Engineering, but once again, there was no certainty that it had led to the illness in question.[20] At the time of writing, while the state of health of children had improved, there was widespread nervousness among the city's population that the disease would return and that those already affected once would suffer repercussions.

Aside from the lamentable plight of the children, the key factor to emerge from the Chernivtsi affair was the ineffectiveness and even incompetence of the official health authorities. One citizen remarked to me sardonically, recalling post-Chernobyl comments, that since (the then) Ukrainian Health Minister Anatolii Romanenko was involved in the inquiry into the alopecia, he would no doubt say that the citizens should not panic, because "the children's hair will grow back even better than before." The statement perhaps typifies the public attitude toward the health authorities, which was one of distrust bordering on contempt. The authorities clearly had not discovered the cause of the illness, indeed had clung to theories even after they had been scientifically disproved (even in 1990, one continued to find references

19. *Robitnycha hazeta*, December 5, 1989; *Radio Kiev*, October 24, 1989. For a related discussion, see *Znannya ta pratsya*, No. 9, 1989, pp. 6–7.
20. See, for example, *News From Ukraine*, No. 30, 1989.

to thallium as the cause of the hair loss). Industrial pollution was claiming victims, even though the government of the USSR had established nature protection committees. In such circumstances, the Green World began to take its place on the Ukrainian scene, as a movement or association that operated independently from any government or party control, and was at the same time not afraid to state the plain truth.

YURII SHCHERBAK

The man who was to emerge as the leading figure in and chairman of Green World was a quiet-spoken family man and former medical practitioner called Yurii Mykolaiovych Shcherbak. In the spring of 1989, Shcherbak ran for the Congress of People's Deputies in the Shevchenko territorial district of Kiev. He was, at best, a long shot, facing six other candidates, a nonparty member and one whose concern for environmental issues had become an embarrassment to the party authorities. As a result, a propaganda campaign was mounted against him. Much was made of the fact that his wife was Polish; that his brother had once been arrested (this was in the Stalin period and he had been completely exonerated since); it was alleged that he had a foreign bank account; that he was a "Zionist" and a "Ukrainian bourgeois nationalist." In short he was regarded as anti-establishment. All the same, he achieved a surprising victory, obtaining about 57 percent of the total vote, and thereafter he was also elected from the Congress to the Supreme Soviet, where he was appointed Chairman of the Subcommittee on Ecology.

In June 1989, it was possible to accompany Shcherbak to a Kiev meeting at the Palace of Culture, at which he and other deputies elected in the Kiev district were called upon to explain a faltering performance at the First Congress. Indeed, the Ukrainian deputies had been virtually invisible. Whereas the first speaker, Valery Hryshchuk, made a variety of excuses for the lack of speeches from Ukrainians or about Ukraine, Shcherbak was more forthcoming. He had in fact published the speech that he was unable to give: a fiery attack on the Chairman of the Ukrainian Council of Ministers, Vitalii Masol, which appeared in the newspaper *Literaturna Ukraina* (June 14, 1989). The voters were unhappy. Several deputies decided not to attend the meeting in such a hostile atmosphere, including the acknowledged leader of the Kiev deputies, Borys Oliinyk. But Shcherbak soldiered on for more than two hours through a battery of questions, even while it was apparent that in general his responses were considerably more moderate in tone than the mood of the audience would have desired. The performance said much about the man who leads the Ukrainian ecological movement.

Yurii Shcherbak was born in Kiev, on October 12, 1934, the year he often notes "after the Ukrainian Famine." He graduated from the Kiev Medical Institute and worked at the Kiev Scientific Research Institute of Epidemiology and Infectious Diseases until 1987. As one of his interviewers has pointed out,[21] Shcherbak has long had a second major interest, however. When he first defended his candidate's dissertation in 1966, his first collection of stories, entitled *As in Wartime,* was in the process of being published. By 1989, Shcherbak had written nine novels before his political duties put a stop to his productivity. In 1984, he was awarded the Yurii Yanovsky Prize for novel writing and, the following year, he received the Dovzhenko Prize for screenwriting. As a doctor, Shcherbak also enjoyed a degree of success. He was awarded the Order of the Red Banner of Labor for his work in combating cholera epidemics in Central Asia and in Ukraine.

The disaster at Chernobyl in April 1986 provided Shcherbak with a mission. He was quick to recognize its significance and that the foreseeable consequences of the disaster were being largely concealed from the Soviet public. He therefore went to the Chernobyl zone and remained there for three months, conducting interviews with eyewitnesses. In the book he wrote about this experience, Shcherbak states:

> A year isn't a particularly long span of time even in the life of a person. But in the course of that year [1986]—no, not a year, but just a few months—we all suddenly matured, grew up by a whole epoch, we became harder and more exacting both toward ourselves and toward those who take responsible decisions, those in whose hands human existence and the fate of nature rest.[22]

21. *Robitnycha hazeta,* April 8, 1989.
22. Yurii Shcherbak, *Chernobyl: A Documentary Story* (London: Macmillan), in association with the Canadian Institute of Ukrainian Studies, 1989, p. 2.

Although Chernobyl may not always remain his chief concern, there seems no doubt that it had been the dominant event in his life to date. He has acknowledged that he is still anxious about the future of the damaged reactor unit and as a deputy one of his avowed goals was to help resolve this question.

Shcherbak's chief interest is in ecological problems. In December 1987, he was chiefly responsible for founding Green World (Zelenyi svit, see below), an organization that was formed from the Committee for Defense of Peace, and whose membership combined scientific workers and creative writers. The ecological movement, in Shcherbak's view, had to be built not on material interests but on the morality of its participants.[23] On these grounds, he has long been opposed to nuclear power, arguing that the country has ample alternatives, such as using the power of the wind and the sun. The USSR, he has pointed out, ranked only sixty-seventh in the world in 1989 in the development of alternative sources of energy. He has denied that he and his fellow writers are seeking a return to the Dark Ages, when they would once again have to write by candlelight. The key problems, he declared, were the excessive use of electricity, the amounts lost in transmission, and the fact that the Soviet energy authorities were so one-track-minded that they could not conceive of alternative sources of power. Why not introduce energy-saving technology on a wide basis, he asked, in order to reduce the huge transmission losses?

His political standpoint, while not so easy to discern, appears to be that of a reforming populist, imbued with a strong sense of patriotism and a continuing belief (at least in the period 1987–90) in the future of the Soviet system under Gorbachev. He has been critical of both the late Leonid Brezhnev and post-Brezhnev periods in Ukraine. But quite often, he has shared platforms and podiums with others who have appeared far more radical than he. Thus at the November 1988 ecological demonstration cited above, speakers such as Shipunov and Makar were more outspoken. Not surprisingly therefore, it was Shcherbak who received the most coverage in the media. He had become officially tolerated (though hardly liked) by the Ukrainian establishment, perhaps also because of a clear alignment between Shcher-

23. *Molod' Ukrainy*, March 14, 1989.

bak and other literary figures in Kiev with the policies of Gorbachev. Ukrainian writers in particular have strong grounds to claim that without them glasnost and perestroika would never have gained even a toehold in Ukraine.

Thus, interestingly, when Gorbachev visited Ukraine in February 1989, he spent a considerable amount of time with politicized writers (Rukh members) such as Ivan Drach and Dmytro Pavlychko, in addition to Shcherbak. The Ukrainian party leader, Volodymyr Shcherbytsky, often appeared to be little more than an embarrassed and uncomfortable spectator at such meetings. At that time, Drach and Shcherbak called openly for Shcherbytsky's removal, a far bolder action in early 1989 than it would have been a year later, when the decline of the Communist Party of Ukraine was more apparent. To Moscow, such voices may not have been unwelcome, because they were manifestations of support for Gorbachev's policies. But Shcherbak at that time was still concerned to work within the existing command structure. He pointed out in the spring of 1989 that:

> It is no secret that under the blanket of democratization and glasnost some people are taking advantage of the faith and inexperience of their listeners. Irresponsible attacks on socialism are being permitted. They are trying to thrust people onto an illegal path. We all have to learn to live in democratic conditions, to educate ourselves in a high political culture. Democracy is incompatible with irresponsibility in words and deeds.[24]

In short, a radical reform of society did not necessarily translate into the overthrow of that society. At first, Shcherbak was critical of the platform of the Popular Movement for Perestroika (Rukh) on the grounds that there were "errors" in the published program. The goal, he stated, should be not to violate existing laws, but to work together in changing them for the benefit of citizens.[25]

On the other hand, Shcherbak can hardly be called a conformist. He has long been working on a book about the Ukrainian Famine, an event which claimed several of his close relatives. His

24. *Robitnycha hazeta,* April 8, 1989.
25. *Molod' Ukrainy,* March 14, 1989.

estimate of the total number of victims, at six to eight million, exceeds those of most Western scholars.[26] Furthermore, he is a Ukrainian patriot. A person without a nationality, he has remarked, is akin to a nomad and is like someone who has lost the struggle of life. Consequently he lent his support to the Shevchenko Ukrainian Language Society, chaired by his friend and poet, Dmytro Pavlychko. The Society, according to Shcherbak, must assist the development of the Ukrainian language and national culture and preserve the traditions of the Ukrainian people. He has persisted in an idealistic identification with "the people." The real value of his work as a politician, he has maintained, has been getting to know people, sensing their feelings, their joys and their problems. "One must be among people," he stated, "and share with them equally in both happiness and grief." He expressed delight that whereas in the past, his readers would approach him warily and say little, for the present they had opened their hearts to him.

As a populist and prominent Ukrainian ecologist, Shcherbak gained a considerable following in Kiev. He received delegations from distant villages, whose inhabitants were concerned about the construction of dangerous factories in the vicinity. He has also been one of the prime contacts for the Western media in Kiev, a city which, unlike Moscow, is not always familiar to journalists. His popularity, above all, has stemmed from his honesty and sincerity (which can be rare qualities in politicians in the Soviet Union, as elsewhere). At the same time, he has appeared to some to be too tentative in his policies. In Kiev in 1989, Shcherbak expressed to me his fear that Ukraine could become a society like Poland in 1980, with riots in the streets and chaos prevailing. In order to forestall such an occurrence, today's Ukrainian politician, he felt, must tread carefully. Because of such diplomacy, Shcherbak and the Green World were able to overcome some of their early opposition. To the authorities, they were far more acceptable than the Ukrainian Helsinki Union, for example, or even the Rukh as a whole (though Green World is part of the Rukh). However, by the time of the March 1990 elections to the Ukrainian Supreme Soviet, the authorities delayed the official registration of the Green World until it was too late to field candidates. It seemed that there was a limit to toleration.

26. *Sobesednik*, No. 49, 1988, p. 12.

THE GREEN WORLD

The Green World (Zelenyi svit) was founded in late 1987 as the coordinator of activities of various informal public ecological associations, which were said to exist in almost every oblast of the republic. It was founded officially by the Ukrainian Committee for the Defense of Peace, under the chairmanship of Shcherbak. The three deputy chairmen were Dmytro Grodzinsky, a biologist and academician; V.G. Sakhaev, a doctor of economics; and Yu. Tkachenko, a film director. The association established itself formally with a draft statute and founding congress in 1989 (both of which are dealt with below), but even in its early months of existence, it adopted clearcut goals. These goals were also cited in the ecological program of the Rukh, and it should be emphasized that there was complete agreement between the Rukh and the Green World on ecological questions. They have been described in detail by one of the Green World's leaders.[27]

The Green World's intention was to realize an "ecologically perfect" mode of life by fighting against the pollution of the environment with chemicals and against the use of chemicals in agriculture. In the area of human rights, it has sought to establish ecological crimes in the shape of violations of environmental laws and misrepresentation of the same in official statements. The main culprits, in the association's view, have been the authorities running nuclear energy and land improvement projects. Every person, in the view of the association, must have the right to healthy life conditions and the proper social conditions. Information on pollution of air, water, the soil, and contamination of food must, in the view of Green World, be open and available to the public. and the association itself has strived to provide such information where possible. People must be educated to think "globally" from early childhood onward, and this signifies that education should be rid of any ideological or militaristic content.

Turning to science and culture, the Green World has supported the creation of mathematical models, ideas and hypotheses on the future development of Ukraine. If there is to be future

---

27. Dmytro Grodzinsky, in his paper on "Ecological Anxieties of the Ukraine," which is cited hereafter.

industrial expansion, in other words, then it must, in the association's view, be based on firm ecological laws, using clean technology and with respect for the existing landscape. In cases of doubt, then the authorities must permit a referendum on each individual project. There must be a respect for the natural environment, for national parks, flora and fauna, and an attempt must be made to recover small rivers and "historical landscapes" of the Dnipro River that have already been adversely affected. Particular attention was paid to agriculture, "the main treasure of Ukraine," with emphasis that those areas blessed with fertile soils should be used exclusively for agricultural needs, most notably the Chernozem lands.

Special focus was placed on nuclear power. The association insisted from the outset that the Chernobyl station must be shut down and that all nuclear power stations in Ukraine should eventually be dismantled. First priority was given to the abandonment of the projects at Chyhyryn and the Crimea. However, the problem was that the republic lacked the technology to carry out the dismantling of existing reactors, rendering the population "prisoners" of the existing reactors. Consequently, the immediate aims of the association were restricted to prevention of any new nuclear power plants on Ukrainian soil. In turn, the entire south of Ukraine was in need of regeneration, not merely because of the nuclear power construction, but also because of irrigation schemes and major power complexes. One can posit that some of these goals were utopian and in general the accent was on the negative aspects of economic development without much suggestion of feasible alternatives (other than the somewhat nebulous "energy saving"). It now remained to establish a formal structure for the Green World.

On April 26, 1989, the organization held a meeting and issued its draft statute, which was published in the press three months later for public discussion.[28] The statute was somewhat confusing because it was laid out in reverse order. It therefore listed the organizational structure toward the end of the document. The preface to the statute stated that Ukraine was on the "threshold" of an ecological crisis and that years of uncontrolled exploitation of Ukraine's riches had led to pollution of the atmosphere, water,

28. *Radyans'ka Ukraina,* July 18, 1989.

and soil, that was ten times the permissible maximum, if not more. Five "ecological disasters" were cited in brief: the ruination of the Dniester River; the Chernobyl tragedy; the unexplained sicknesses among the children of Chernivtsi; the contamination of the Dnipro River; and the salination of the soil in southern regions of the republic.

Green World was described as a voluntary association of ecological cells and collective members. The highest body within the association was to be the republican congress, which had to be held every two years under normal circumstances, but could also be called exceptionally with the agreement of one-third of the cells. Representatives of all the cells could be delegates to the congress. The congress was to adopt the statute formally, and make amendments to it, elect the chairman and deputy chairman, and hear the reports of the executive authority. The latter was made up of a Green Council and a Control and Inspection Committee that was appointed by the council. The Green Council, which was to meet four times annually, was to be composed of the chairman and deputies, members of the secretariat, representatives of collective members of the Green World, and experts and scholars. Its task was to examine ecological problems, to represent the association in state and public organizations of Ukraine and the Soviet Union, to develop contacts with foreign ecological organizations, to produce a periodical, and to accept collective members into the association. The day-to-day business of the association was to be run by the secretariat, which again included the chairman and deputy chairmen.

Green World originally had described itself as an informal association, so the question arises why it decided to issue a formal statute after all. The answer appears to lie in the amount of pressure exerted by the authorities against its members for drawing attention to ecological problems. That this might well have been the case is indicated by a clause in the draft statute which stated that it was the duty of the secretariat to provide "members with protection from various forms of victimization and persecution [that result] from their nature conservation activities." It seems likely that such victimization must have taken place in the past. Further evidence of animosity between factory managers and ecological "watchdogs" was suggested under the heading "General Tasks," one of which was to prosecute those individuals or enterprises that violated pollution standards and caused harm to the environment.

One of the key aims was to coordinate public control over industrial pollution and environmental norms. Although Green World was founded, as we have stated, by a "peace committee," there was only a brief allusion in the statute to what might once have been considered a fundamental issue: the danger of nuclear war.[29] There was no membership fee, and local branches were permitted their own internal structure and organization, and financial independence. Funds were to be provided by voluntary contributions of collective and individual members, sponsorship, income from benefit concerts and lectures, and the sale of such goods as posters and badges bearing the emblem of the association. The use of such emblems was to be monitored by the secretariat. The association was to have its own bank account and letterhead, and the official address of Zelenyi svit was published in the press for the first time in the draft statute.

THE FOUNDING CONGRESS

On October 28–29, 1989, Green World held its founding congress in Kiev, an event that was attended by several international organizations, including Greenpeace. A variety of speeches were given, many of which were very emotional. Altogether they painted a devastating picture of the ecological situation in Ukraine, highlighting some of the major ecological hazards afflicting the republic and relating them to the incidence of disease and birth and death rates. The major area of concern was declared to be the Donetsk-Dnipro region, which constitutes the industrial heartland of Ukraine, and encompasses the Dnipropetrovsk, Donetsk, Zaporizhzhya and Kirovohrad oblasts.

29. The brevity is truly significant. After Chernobyl, Mikhail Gorbachev maintained on Soviet television (May 14, 1986) that the event provided justification for his policy of removing all nuclear weapons from the face of the earth by the year 2000, and the significance of Chernobyl in the founding of Zelenyi svit has been made clear. Similarly, in their book on Chernobyl, *Final Warning: The Legacy of Chernobyl* (Warner Books, 1988), Robert Peter Gale and Thomas Hauser devoted the entire last part (forty pages in all) to this same topic. One can posit that by 1989, members of Zelenyi svit had recognized that Chernobyl offered a more important message about the state of the existing environment, in Ukraine in particular.

Our analysis will focus on the main speech delivered by Shcherbak.

Shcherbak began by listing some of the successes achieved by the association in its almost two years of existence. It had played a major part, he stated, in the stoppage of work on the Danube-Dnipro Canal, on the third "stage" of the Chernobyl nuclear power plant (reactors five and six), on the Odessa, Chyhyryn and Crimean nuclear power plants, on unit four of the South Ukraine nuclear plant, and on a chemical combine in the Crimea. He pointed out, however, that these triumphs had been dwarfed by the appalling degradation of the natural environment in the Donetsk-Dnipro region, which was by this time the most polluted region in the Soviet Union.[30] These observations can be supported, incidentally, by other evidence. Thus in an article published in late 1989, V. Popov, the First Deputy Chairman of the Ukrainian State Planning Committee, maintained that despite a slight reduction in the number of harmful byproducts released into the atmosphere in the first half of 1989 as compared with the same period in 1988, the level of atmospheric contamination remained "impermissibly high."[31]

Shcherbak also stated that it was "our national and social dis grace" that life expectancy in Ukraine for men was seven to eight years less and for women four to six years less than that in developed countries. The republic had a high rate of "mutilated" births—six to thirteen per 1,000—and also the lowest birth rate in the Soviet Union, with a tendency for the rate to decrease further. The Soviet Union as whole, he declared, was in a lowly twentieth place worldwide for the longevity of human existence. Women made up 80 percent of those involved in heavy physical labor, and this had led to a rise in illnesses during pregnancy and to a four- to six-fold increase in the number of miscarriages in recent years. On average, Shcherbak noted, in Ukraine, 40,000 pregnancies did not run to full term. These health problems could be related directly, in his view, to the ecological damage caused by heavy industry.

In such polluted areas as Zaporizhzhya, Dniprodzerzhinsk, Rubizhne and Kremenchuh, the incidence of oncological diseases

---

30. *Literaturna Ukraina*, December 14, 1989.
31. V. Popov, "Pro khid sotsial'no-ekonomichnoho rozvytku Ukrains'koi RSR u chetvertomu rotsi dvanadtsyatoi p'yatyrichky," *Ekonomika Radyans'koi*

among children was five to eight times the national average, and the highest number of deaths from cancer also occurred in these areas, Shcherbak declared. Cities like Zaporizhzhya (the subject of a gruesome documentary film about birth defects in new babies called "Hostages"), Dnipropetrovsk, Odessa, Cherkasy and Severodonetsk could already be called "zones of economic calamity." Once again, further evidence appeared at this same time that corroborated Shcherbak's startling announcements about the relationship between environmental pollution and physical illnesses, especially among newborn babies and young children.

For example, an article by Grigorii Shmatkov, the chief ecologist at the Ukrainian Academy of Sciences Dnipro Research Center, substantiated such statements. Shmatkov wrote that the Center tested a group of mothers breastfeeding new babies and discovered that 80 percent of them were producing milk that contained chemicals harmful to their children. "It hurts to look at those kids," he remarked.[32] Shmatkov went on to say that the Dnipro Research Center had sent a commission to the Petrovsky steel mill in Dnipropetrovsk, an old enterprise that "spews out tons of toxic gases" (see also above). Both there and at the nearby Dnipro thermal electric station, the air was said to be so poisonous that "the human organism cannot adapt to it" and the circulatory and muscular systems were adversely affected. Birth rates in Dnipropetrovsk had not risen for a decade, but the mortality rate had risen continuously. The Dnipro region had acted as an employment magnet, emptying villages of their populations for industrial needs and then subjecting these workers to pollution levels that significantly increased the incidence of high blood pressure, cancer and other diseases among them.[33]

Returning to the speech by Shcherbak, he mounted a strong attack on what he perceived as the archaic structure of Ukrainian industry, which had remained in place despite the enormity of the environmental concerns. A change in thinking had to come, he felt, first from the Soviet government and second from the Ukrainian government. However, in his view, the only real solution was the granting of economic sovereignty to Ukraine within

---

Ukrainy, No. 10, 1989, p. 7.
32. *Ukraine*, No. 11, 1989, p. 11.
33. *Ibid.*, p. 12. See also *Komsomol'skaya znamya*, September 16, 1989.

a reformed federation of Soviet republics.[34] This would end once and for all what Shcherbak called the hegemony of the colonialistic, monopolistic and rapacious all-Union ministries and associations that had been the main cause of the current ecological dilemma of the republic. Shcherbak declared that the right of citizens to live in an ecologically safe environment should be made an article of the Ukrainian Constitution.

In connection with the Chernobyl disaster, Shcherbak assailed the doctors who had "lied" for three years about the impact of this tragedy, a reference to the Ukrainian Ministry of Health and All-Union Center for Radiation Medicine.[35] He demanded the punishment of those who had built graphite-moderated reactors in the Soviet Union, and of those who had failed to take appropriate action in the first hours and days after the accident. He described Chernobyl as representing not so much a scenario that might occur in a nuclear war, as a case of warfare in peacetime waged by the Ministry of Nuclear Power Generation and the Nuclear Industry (formerly the Ministry of Nuclear Power) against its own people. Zelenyi svit, he suggested, should compile a "Black Book" on Chernobyl, as the first truthful documentation of the accident, and the book should be published in several languages, including English. He called for a ten-year moratorium on building nuclear power plants in the Soviet Union, and for no further investment in existing plants.

But where did Zelenyi svit stand politically? On this question, Shcherbak was more evasive. He declared to the assembly that the association was prepared to cooperate with anyone concerned about the fate of Ukraine, including the Ukrainian Communist Party, the Rukh, Ukrainian government and nongovernment organizations, the Russian Orthodox, Ukrainian Catholic and Roman Catholic Churches, and the Jewish, Turkic and Buddhist communities, because "ecology must be placed above eco-

34. This might be described as the minimalist program of Ukrainian reformers in the period 1989–90, and was advocated, inter alia, by the Communist Party itself. The maximalist program was for immediate political and economic independence of Ukraine, expressed by the Ukrainian Republican Party, founded in May 1990. See *Rabochaya trybuna*, May 12, 1990.
35. Surprisingly, the official secrecy was finally acknowledged by a spokesperson from the Center in the late spring of 1990, who stated that information about the medical effects of the disaster had been classified. *News From Ukraine*, No. 23, June 1990.

nomics, above all political dogma and myths." Having made this sweeping statement, Shcherbak initially wavered on the question of creating a separate Green Party, and decided ultimately to leave the decision to the delegates.

An account in *Radyanska Ukraina* noted that there were three possible paths for the association to take. The first was to become a voluntary association of various groups united in the struggle to avert an ecological crisis. The second was to become a branch of the world-wide Greenpeace organization. The third was to form a Green Party of Ukraine similar to the successful Green Party in the Federal Republic of Germany.[36] Shcherbak supported the idea of a Green Party that could operate in conjunction with the Communist Party of Ukraine; its members would be able to hold dual membership of both parties because the Green Party would espouse no ideological doctrine and its platform would be dictated purely by ecological issues. However, some delegates went further. Andrii Hlazovy, a journalist, declared:

> The fact of the matter is that we feel the time has come to set up a party on the basis of the ideology common to all the Greens in the various countries of the world in order to work more effectively for implementation of our principles in everyday life.[37]

In short, Hlazovy was advocating the formation of a Green Party that could immediately begin working on a platform for the spring 1990 elections, though this opportunity was to be denied to the Greens, as noted above.

The most acute problem facing the association in the wake of the congress was the impossibility of separating ecological and political issues. While the Green World had achieved some success, it could be argued that its power and influence would be greatly enhanced if it were a legal political party. Shcherbak had in fact been elected to the Congress of People's Deputies in the spring of 1989 on a predominantly ecological platform. By its attacks on the bureaucracy and on the nature of industrial

36. *Radyans'ka Ukraina*, November 15, 1989.
37. *News From Ukraine*, No. 45, November 1989.

decision-making in Ukraine, and by its unceasing opposition to nuclear power plants, Zelenyi svit had ipso facto adopted a political stance that divorced it from contemporary party ideology. Even Lenin had stressed the importance of electricity in a future Soviet economy, and now Zelenyi svit at the least was promoting a policy that would reduce the electricity generating capacity of Ukraine. Thus even in late 1989, it seemed that Shcherbak's statement that it would be possible to hold simultaneous memberships in the party and the Green World had already been negated. However, before the issue of a Green Party was to be finally resolved, there followed an interlude, during which the Ukrainian Supreme Soviet decided, belatedly, to hold a session devoted to the ecological situation in Ukraine.

THE UKRAINIAN SUPREME SOVIET MEETING, FEBRUARY 1990

In late February 1990, the Ukrainian Supreme Soviet[38] adopted a decree "Concerning the Ecological Situation in Ukraine and Measures for Its Radical Improvement."[39] The decree followed several days of animated and often bitter discussion about Ukraine's environment. It indicated that the authorities were concerned about the situation and wished to be seen to be taking steps to address the most serious problems. Yet its clauses were often vague and nebulous. Moreover, much reliance was placed in the decree on the Ukrainian State Committee for the Protection of Nature, an organization not known hitherto for being either aggressive or effective in its demands. The decree began with a lengthy preamble that described part of the republic as being "on the edge of an ecological crisis," including the Dnipro region, the Donbass, the Krivyi Rih area, the northern Crimea, and the Black Sea and Sea of Azov. A third of all Ukraine's arable land, it noted, was suffering from soil erosion, while chemical substances continued to contaminate it and further deplete its value. In 1989, about seventeen million tons of poisonous substances were released into the atmosphere, about one-third of it by motor transport; twenty Ukrainian cities were said to be highly contaminated.

---

38. It should be recalled that reference here is to the preelection Supreme Soviet, i.e., there was not yet an official opposition, and many of the more "conservative" representatives were still in office.
39. *Radyans'ka Ukraina*, March 1, 1990.

The Ukrainian Supreme Soviet blamed the Ukrainian Council of Ministers for its lack of vigilance and failure to use its full powers to alleviate the situation and also indicted all-Union organizations and ministries for violations of ecological laws. In consequence, the major clause of the new decree required that a plan for the future economic development of Ukraine be drawn up with proper regard for the environment that anticipated a sharp reduction in further development of raw materials extraction, energy production, and industries that affected Ukraine's water supply. All new industries that would not directly benefit the public were to be halted in major cities and emphasis placed instead on the fulfilment of schemes to protect nature. Economic methods were to replace what the Supreme Soviet called "administrative-command" methods of control in industry, and enterprises that violated antipollution regulations were to be fined or made to relinquish their profits.

The Council of Ministers, oblast executive committees and the city councils of Kiev and Sevastopol (the Crimea had, increasingly, been treated as a separate entity by both the Ukrainian party authorities and the Rukh after 1989) were requested to draw up a state-wide program for the protection of the environment and for the rational use of natural resources in Ukraine by 1991. By 1995, the volume of dangerous substances released by large factories was to be reduced to permitted levels. Facilities for the purification of water were to be constructed in all of the republic's cities by 1996, and the flow of poisonous products into Ukraine's major rivers, the Dnipro, Dniester, Desna and South Buh, was to be stopped completely by the year 2000. The statements sounded comprehensive enough, though one should bear in mind that in theory there were already laws in place that should have stopped indiscriminate pollution of the environment. Even maximum norms had been established for pollution levels. The main problem was that because economic needs had been declared paramount since the time of Stalin, then such regulations were more often than not ignored by planners.

A third clause in the decree related to Chernobyl and the need to complete urgent evacuations in 1990–91. In addition, the Ukrainian government had resolved to start up its own "Children of Chernobyl" program (to match the one in the West), the purpose of which was to supervise rapid evacuation of children under fourteen and all pregnant women from contaminated centers by the end of 1990. Over a six-month period, the Ukrainian government and USSR Ministry of Nuclear Power and Nuclear Industry were requested to elaborate plans for the shutdown of the Chernobyl nuclear power plant by 1995, and the government was also asked to resolve questions concerning the halting of new reactors at the Rivne and Khmelnytsky stations and the shutdown of an experimental reactor operated by the Ukrainian Academy of Sciences in Kiev.

The decree required enterprises and factories to be examined thoroughly in 1990–91 to ascertain whether they posed a danger to the environment, and the Ukrainian State Agroindustrial Committee was requested to supervise the provision of uncontaminated and varied food for children. Ukrainian research institutes were asked to investigate further the effects of radiation on the everyday life of the public and to assess what agricultural activities could be conducted in areas where radiation levels remained high. The Ukrainian government was advised to draft several new laws by 1991, including a Law of the Ukrainian SSR for the Protection of Nature, a Law for the Protection of Nature Reserves, and a Law about Atomic Energy and Atomic Safety. Perhaps more important, the Ukrainian Supreme Soviet decree demanded that the government transfer institutions subordinate to the so-called Third Department of the USSR Ministry of Health located in Ukraine to the jurisdiction of the Ukrainian Ministry of Health without loss of its present facilities. This decision was a direct result of pressure from the Green World, and Shcherbak in particular.

In other spheres, however, the Supreme Soviet was obliged to turn directly to all-Union ministries. For example, the USSR Ministry of Power and Electrification was asked to provide equipment during the course of the 1991–95 plan period to scrub acid-forming substances from fumes. Centralized capital investment was also needed to enable the USSR Ministries of Ferrous Metallurgy and Coal to treat highly mineralized waters in the western part of the Donbass and the Krivyi Rih area. The USSR State Construction Committee, according to the decree, had to arrange for the burial of toxic industrial wastes on Ukrainian territory. In some ways, it was humbling for the Ukrainian parliament to be forced to place such reliance on ministries outside the republic. It was also unclear why this should have been the case. On January 1, 1991, economic sovereignty was to be introduced into Ukraine, and thus in theory, the control over these industrial polluters was to pass to the republican government.[40] One possible explanation for such apparent timidity is that the deputies remained uncertain of whether Ukraine would indeed gain control over such strongly centralized industries as coal, steel and nuclear power.

There are other criticisms that can be made of the Supreme Soviet decree. On Chernobyl, for example, there was no single program, but several separate points, many of which amounted to bureaucratic rhetoric. The decree made reference to the future role of the State Committee for the Protection of Nature, but not to Green World, the activity of which had compelled the parliament to discuss ecological issues in the first place. The Supreme Soviet, in addition, was in its "last throes" before a new election took place, so there was a suspicion that the whole discussion was somewhat ritualistic. Yet the mood of the republic had been made evident, and even the arguments advanced by the supporters of nuclear power—such as Mikhail Umanets, Director of the Chernobyl plant, for example—were subdued.

ANALYSIS

On April 26, 1990, the fourth anniversary of Chernobyl, the Green Party was officially founded in Ukraine, as a political party alongside (not replacing) the Green World association. Thirty members of Green World, including Shcherbak, signed the resolution to establish a Green Party, and its draft program was released on March 23. The program singled out nuclear energy, not only as an undesirable industry because of its danger to the ecology, but also because of its alleged administrative-command structure (a term used mainly to describe the organization of the economy in the Brezhnev period), whereby all decisions on plan-

---

40. A number of draft plans for the economic sovereignty of Ukraine had already been discussed, as, for example, on Radio Kiev on November 21, 1989. The initial draft was published in *Radyans'ka Ukraina*, August 6, 1989.

ning, location and operation of stations have been made in Moscow. The Green Party program advocated the outright prohibition of nuclear power, and demanded the adoption of a new law on energy and energetics, whereby two-thirds of the funds allotted by the state for energy needs would be used for the development of renewed and alternative energy sources.[41]

The current prominence of the Green World and Green Party arose from official hostility or apathy toward public concerns over the environment. Shcherbak once conveyed an image of a typical bureaucrat of the Brezhnev era, sitting in a comfortable armchair in Moscow and making decisions about Ukrainian energy with the stroke of a pen, and certainly without examining or visiting the area in question. Phrases such as "the period of stagnation" have been used frequently by Soviet analysts to describe the Brezhnev period. They are not altogether accurate because they refer to the post-1975 period rather than the first decade of Brezhnev's leadership. In fact the early years might even be called "years of reform," commencing with the Brezhnev-Kosygin economic reform of 1965–66. It is fair to say, however, that economic development in this time and beyond paid little or no attention to ecological concerns. Grandiose schemes were drawn up that were often praised for their sweeping nature: diverting Arctic waters to Central Asia; cutting off the Dnipro River from the Black Sea; and others. It was Stalin who had introduced such expansiveness and these schemes were part of the makeup of the Soviet regime.

The Ukrainian case may not have been worse than other parts of the Soviet Union (or even the heavily polluted countries of Eastern Europe, such as Poland and East Germany). It was the extent and thoughtlessness of the industrial build up that had begun to disturb people, even before Chernobyl. Vacation areas, such as the Crimea and Odessa had become industrial eyesores. The older industrial regions—the cities of Eastern Ukraine—were in a dire state and particularly neglected because of the passivity of the population, cowed by a repressive regime that at times exceeded any other regional party organization in its ferocity during the Stalin years. Both glasnost and the individual ecological

41. *Zelenyi svit*, No. 1, April 1990.

catastrophe of Chernobyl were needed to undermine this old system, but even then it clung to its old power and traditional values. "Experts" were called upon to denounce this "emotional posturing" of the public that in their view had been incited by irresponsible journalists. Even at the very top there was no clear path ahead. Gorbachev was to be found in 1985 praising the concept of Stakhanovism that had been partly responsible for the attitude of "production for the sake of production itself."

Yet there was a path ahead. There is a contrast between movements, such as the various peace committees organized by Stalin, which began at the top and those like Zelenyi svit, a grassroots ecological movement that relied on public anger in order to develop. To some extent it was spontaneous. Shcherbak, Grodzinsky and others harnessed this public mood by keeping the movement as loose as possible. The move toward a Green Party arose only because of the refusal of the giant dinosaur that the Ukrainian Communist Party had become to tolerate such an association during its first two years of existence. Zelenyi svit became popular, chiefly, because the public could perceive, after Chernobyl, how their land was being systematically destroyed in the name of economic progress. The movement grew because its leaders, through the media of the press, television and radio (not through academic journals, which had fallen well behind the press in terms of openness and progressive thinking), convinced the people that such degradation might be irrevocable. In addition, the official secrecy over events such as Chernobyl—the supreme irony was perhaps the West's perception that Chernobyl was the first evidence of glasnost on the part of the Soviet regime—suggested that even under an enlightened leader, little had really changed.

Finally, by 1989–90, the Ukrainian republic had been pervaded by outbursts of patriotism and national feeling. The blue and yellow flag had replaced the state flag of the Ukrainian SSR, even in many parts of Kiev; the emblem of the tryzub (three-cornered star) had been resurrected. Such manifestations were led by Western Ukraine, annexed by Stalin only during the Second World War, and then the location of a bitter guerrilla warfare against the Stalin regime until the early 1950s. Ecological decay was one issue on which all Ukrainians and all those living in

Ukraine could be united: anyone who had visited a city like Zaporizhzhya or witnessed the chemical poisoning of coal miners in Horlivka[42] did not need to be informed that here was a problem that could not be left to fester or that very little was being done to alleviate the situation at the top—in this case, in Moscow.

42. *Radyans'ka Ukraina*, June 19, 1990; *News From Ukraine*, No. 24, June 1990. See also the account of the pollution of Voroshilovhrad (Luhanske) in *Ekonomika i zhizn'*, No. 12, 1990.

6 | THE DONBASS MINERS AND THE 1989 COAL STRIKE

THE DECLINE OF UKRAINIAN industry is perhaps typified by the serious problems in the coal basins, and in particular by the systemic and deep-rooted dilemmas associated with the Don River Basin (or Donbass). In addition, the situation in the Ukrainian coal mines has provided another example of a public response to an intolerable situation, this time in the shape of the first major strike in Soviet history, which took place in July 1989. An interesting facet of the coal strike—in its Ukrainian manifestation, for it actually began in the coal mines of Siberia— was that militant action was being undertaken by the least nationally conscious group of people in the republic. Here, there was no tradition of nationalism or revolt. Quite the contrary. East Ukrainian workers have been almost notoriously passive, kept in check by an extremely repressive party organization which rendered negligible any valid trade unions that might work to support the workers' interests. The coal miner, it was always felt, lay steeped in the Stakhanovite tradition, virtually unaffected by perestroika, and certainly unlikely to offer an overt protest. Such an attitude toward the miner could hardly have been more misguided. This chapter will examine the background to the strike, the events themselves, and place them within the context of the outburst of public opinion in Ukraine that has arisen from a worsening ecological and (in this case) economic situation.

THE STAKHANOVITE TRADITION

Aleksei G. Stakhanov was perhaps an unlikely Soviet hero, a simple Donbass miner who, under conditions that were very much artificially created, hewed 102 tons of coal in five hours and forty-five minutes on the night of August 31, 1935 at the coalface of the mine "Tsentralnaya-Irmino." The rate exceeded the established norm by fourteen times and inaugurated the Stakhanovite movement which spread across the Soviet Union at the behest of the Stalin regime. In the mines, Stakhanov's record was quickly surpassed: a party member in a neighboring mine, Miron Dyukanov, hewed 115 tons of coal on September 4, 1935; a young miner called Dmitrii Kontsedalov surpassed this total by ten tons on the following day. Not to be outdone, Stakhanov returned to the fray, with a remarkable total of 175 tons in a shift, to be outdone by one Nikita Izotov, with 241 tons per shift later in this same month.[1] Accomplishment was based on sheer quantity, on the selfless sacrifice of the worker for the good of the state, and while one could commend such attitudes, they spelled the beginning of the problems in the Donbass coalfield in particular.

During the process, safety needs were neglected, mines were exploited carelessly and the effects of such work norms on the health of the miners were drastic indeed. The period was one of massive industrialization, and in the coal industry it was based largely on the Donbass coalfield (or Donetsk Basin), which stretches into Rostov Oblast of the Russian Republic, though about 85 percent of the reserves lie on the territory of Ukraine. Record output became the norm, but what was perhaps surprising is that the period was venerated also during Gorbachev's first year as General Secretary of the CC CPSU, which coincided with the fiftieth anniversary of Stakhanov's feat. It was a year in which Gorbachev was formulating his policies for the future, though his main directions, other than the anti-alcohol campaign launched during this year, were unclear. For the miners of the Donbass, it appeared to be the case that the Gorbachev era would not herald any kind of improvement in their situation (the basics of which are portrayed below).

1. *Ugol' Ukrainy*, No. 8, August 1985. Stakhanov died in 1977, and the town of Kadievka in the Donetsk Basin was renamed Stakhanov in his honor.

According to *Pravda,* in a fiftieth anniversary article, Stakhanovism's importance lay in more than the physical feat itself. It had overturned the usual notion about the production potential of the "people," and carried along in its wake millions of people which, in turn, had enabled a dramatic progress in the development of the Soviet economy. Stakhanovism had sparked socialist competition, shock-labor, the "subbotniki," whereby workers would give up their Saturdays and provide a day of free work. Comparisons were made between "Stakhanovite collectivism" and the contemporary organization of labor in "khozraschet" brigades. It was pointed out that several enterprises had once again embarked in the spirit of Stakhanov by raising their production norms. While stress was on quantitative output—numerous examples were provided in the article, especially from the coal industry—the new desire for what was termed "scientific-technical progress" was also emphasized.[2] It seemed as though the new era had at once reverted to the old.

On the anniversary of Stakhanov's feat, fifteen miners at his former coal mine (the name of which had been changed from "Tsentralnaya-Irmino" to "Twenty-Second CPSU Congress") exceeded the original record by fulfilling from fifteen to twenty work norms during a shift and had extracted more than 22,000 tons of coking coal in excess of the plan since the start of 1985. A letter was sent to Gorbachev, who replied as follows:

At that time [1935], our heroic working class, which had mastered equipment that was advanced for the times, made a breakthrough, literally, along the entire front of scientific and technical progress, This secured the economic independence of the country and made it possible to create a reliable defensive shield, against which the Fascist aggressor smashed his head.... There are great lessons in the Stakhanovite movement. They have permanent significance today when we are faced with the task of putting into practice just such a massive, but on a far greater scale, labor breakthrough in the sphere of the intensification of the national economy, accelera-

2. *Pravda,* August 31, 1985.

tion of scientific and technical progress and a restructuring of thinking of all levels of staff.[3]

On the anniversary date itself, the Stakhanovite traditions were celebrated with great fervor and featured throughout the Soviet media. Further, a group of veterans of the movement went to Moscow and met with Gorbachev, who referred to them as the "living history of our homeland." In his speech at the end of the meeting, Gorbachev described the mid-1930s as a time when it was necessary to make a dramatic leap forward in industry, and today, because of the Stakhanovites, it was necessary to thank those who had been pioneers of the movement. However, the lessons for the present were also made clear by the Soviet leader. This style of movement would always form the basis for the resolution of new tasks by the party. Today, he noted, the tremendous potential of the Soviet man must again be harnessed, using scientific knowledge and the hard work and patriotism of the working class and the peasantry.[4] The message was clear. A new and vigorous leader had exhorted the miners to continue the old traditions.

THE DONETSK BASIN VERSUS MOSCOW, 1985–88

The problem was that even if one assumes that Gorbachev was wholly serious—and while he probably could not have ignored the anniversary, there was no necessity for him to take such an active part in it, so it appears that his enthusiasm was genuine—the conditions in the mid-1980s and the mid-1930s were quite different. The coal mines of the Donbass were in a dilapidated state, not least because of the famed Stakhanovite traditions. Easily available coal had been extracted so that what remained in the 121 coal mines lay in thin and sloping seams, often more than 1,200 meters underground. One mine in the Torez region was reported to be 1,319 meters deep in 1985, which was a new record depth for the Soviet Union.[5] The situation was becoming worse with each passing year as the average depth of the coal-

3. Radio Moscow, August 28, 1985.
4. Radio Moscow, September 21, 1985.
5. *Visti z Ukrainy*, No. 30, July 1985.

faces increased by 10–15 meters. In addition the quality of the coal was declining. In the Voroshilovhrad Coal Association, for example, the average ash content of the coal had risen by 6.3 percent over the period 1971–85.[6]

At the start of the Gorbachev period, the Ukrainian coal mines lacked the equipment to alleviate their predicament of falling output and complicated geological conditions. In the 1981–85 plan period, labor productivity in the Soviet coal mines as a whole fell by 6.5 percent, according to the then Minister of the Coal Industry of the USSR, Borys Bratchenko. In a report to a conference of party and economic workers of the Coal Ministry, in late July 1985, he noted that:

> Every fourth face in the coal mines of the Ministry of the Coal Industry of the Ukrainian SSR provides less than 300 tons of coal daily. In the Podmoskovskii Basin, 27 percent of them give less than 500 tons per day, and in the Kuznetsk Basin, the figure is 15 percent of all coal faces. This is a result of the low technical level of machinery, the ineffective organization of mining work, a shortage of qualified cadres, [and] difficulties encountered in trying to acquire spare parts of machinery.[7]

Fully 75 percent of all jobs in the coal miners were being carried out manually. Bratchenko stated that 40 percent of scientific-research and planning-construction institutes were not fulfilling their potential. In the Donetsk Basin, only 12 percent of coal mines possessed the standard equipment to clear coal faces. Furthermore, when new machines were introduced, often no one knew how to use them. At the "Zarya" coal mine in the Torez district, for example, as Bratchenko had noted in this same report, a new machine had been brought to the mine. But because neither the constructors nor the miners were "equal to the task of operating it," the motor for the machine had to be changed three times within six months as a result of ill usage.

A further problem for Ukraine was the declining share of output in the all-Union total. The total coal output in 1970 had been 207.1 million tons, by 1975, it had risen to 222 million

---

6. *Robitnycha hazeta,* January 26, 1986.
7. *Sotsialisticheskaya industriya,* July 25, 1985.

tons, but it had declined to 189 million tons by 1985.[8] Ukraine's share of the total output, which had been over 50 percent in the early postwar years, fell to 26.5 percent by 1985, and had dropped further to 24.9 percent by the year 1988. Under these circumstances, the Soviet authorities decided that it was sensible to reduce investment into the old coalfield and to raise that in the Kuznetsk Basin of Siberia in particular. In order to take such a step, the planners in Moscow had to overcome furious opposition from the Ukrainian Ministry of the Coal Industry which, after years of passivity, showed surprising passion in defending its territory, and in arguing that the Donetsk Basin had a long and viable future as a coalfield. This argument gained plausibility after the Chernobyl disaster when it became apparent that the high hopes placed in nuclear energy as a guaranteed source of power had been unwarranted. In several articles and addresses, Mykola Hrynko, the Ukrainian Minister of the Coal Industry, argued the case for the old coalfield.[9]

At the same time, however, there were accusations forthcoming that Hrynko was unable to keep his own house in order. An instructive example of the state of affairs in the Ukrainian coal mines in 1985 came in the shape of an open letter to the minister, which was published in the newspaper *Pravda Ukrainy*.[10] There is a possibility that the appearance of the letter was part of a campaign, engineered in Moscow, to remove the troublesome Hrynko. If so, however, then it must also be acknowledged that the difficulties described in the letter were genuine enough, and some of them were to reappear in the complaints at the time of the 1989 strike. Thus they form a prelude to the crisis that was to emerge and signalled also the death knell not only of Hrynko's tenure, but of the Ukrainian Coal Ministry as a whole.

The letter was authored by a brigade leader and a squad leader from the "Komsomolets Donbassa" coal mine, which had come into operation at the end of 1980 and was reported to be the largest and most modern in the Soviet coal industry. Its daily capacity was declared initially to be 7,000 tons of fuel. Accord-

---

8. *Narodne hospodarstvo Ukrains'koi RSR u 1987 rotsi: Statystychnyi shchorichnyk.* Kiev, 1988, p. 77; *Soviet Geography*, April 1985, pp. 301–2.
9. The most important and well argued article appeared in *Izvestiya*, December 4, 1984.
10. *Pravda Ukrainy*, September 24, 1985.

ing to the two authors of the letter, however, it had yet to exceed a daily output of 5,000 tons. In fact, they declared, the Five-Year Plan had broken down for all indices, principally because of the "superficial style of leadership" at the coal mine. Instead of providing "concrete help," workers from the Ukrainian Ministry of the Coal Industry had paid a series of ritualistic flying visits to the mine. In April 1985, Hrynko himself had made a trip there and held a conference that was attended by workers of all ranks.

The conference had put forward several concrete reasons why the mine was failing to meet the demands of the plan: incorrect sinking techniques, low quality of work, insufficiently qualified mechanics, and others. Hrynko had counselled the mine leaders to make an agreement with a local machine construction works so that the mine could receive better equipment. But this and other resolutions, according to the authors, were simply "paper decrees." In the somewhat strained metaphor of the writers, the officials from the Ukrainian ministry were talking with "honey in their mouths" but actually producing "the bitterest of peppers." In August, another group of specialists, including technicians and scientists prominent in their fields, visited the mine to ascertain the reasons for the unsatisfactory progress. Evidently they had merely reiterated what was already well known: that the machine being used was unsuitable for one of the seams.

The writers lambasted these "show activities" while stressing that they were all being carried out in Hrynko's name. Committees were said to be "swamping" the mine. Ten to fifteen people reportedly arrived at the mine every day to check up on various aspects of work, which diverted mine leaders from important mining operations. Plans were being constantly changed. In sections 1 and 6, for example, the plan for fuel output had been changed every month in 1985, which had had a demoralizing effect on the workers. Nor was the leadership of the mine very stable. In less than five years of operation, the mine had had three directors, three chief engineers, and innumerable changes of section leaders. The feeling of uncertainty had been heightened, said the authors, by the inept selections for mine director. The first choice was a man who was "morally unscrupulous"; the second too authoritarian and unable to communicate with the miners; while the third and present director was inaccessible to the average worker.

The letter concluded by demanding a response from Hrynko to these problems within the leadership. The tone was far from respectful and the implication was that the Ukrainian Coal Minister was incompetent. Given the structure of the Soviet coal industry, and the hitherto ruthless suppression of any manifestations of independent thinking,[11] it seems likely that the move against Hrynko was organized from a higher level. In any event, the letter and complaints gave the authorities an excuse to remove Hrynko, and on October 23, 1985, he was "released from his duties in connection with his transfer to other work" (that is, fired) and replaced by Nikolai Surgai, his first deputy.[12] Hrynko's removal had significance beyond the sphere of the coal industry, in that he was the first major figure to be removed from the Ukrainian hierarchy in the Gorbachev period. This was partly because he had become troublesome—with an at least quasi-patriotic attachment to the old coalfield—and partly because the Ukrainian coal industry had been designated for widescale reform, and Nikolai Surgai appeared, initially, to be somewhat more pliable (though this did not turn out to be the case).

Some twenty months after Hrynko's removal, the Ukrainian Ministry of the Coal Industry (along with the Ministry of Ferrous Metallurgy) was abolished as a prelude to the restructuring of the entire industry. In the summer of 1987, state production associations were created in Donetsk and Voroshilovhrad oblasts to replace the former ministry. On September 16, 1987, Surgai and V.I. Poltavets were selected as directors of the new production associations, portfolios, one might add, that were intended to be without much authority since the idea was that the industry would become independent. The unfortunate Surgai had had little time to have much influence on the Ukrainian coal industry, but from 1975 to 1982 he had been Director of the old Donetsk production association, so was back in familiar territory. Poltavets was a fifty-one-year-old mining engineer and graduate of the Dnipropetrovsk Mining Institute who had been appointed Deputy Minister of the Ukrainian Coal Industry in 1986.[13]

11. An attempt to establish an independent trade union, for example, had resulted in the incarceration of the initiators in psychiatric hospitals. See, for example, Bohdan Nahaylo, "The Death of Soviet Workers' Rights Activist Aleksei Nikitin," *Radio Liberty Research Bulletin*, RL 166/84, April 25, 1984.
12. Radio Kiev, October 23, 1985.
13. *Ugol' Ukrainy*, No. 12, 1987, p. 2.

The two new associations were transferred, as of January 1, 1988, to "full self-accounting and self-financing." Officially they were now subordinated directly to the USSR Ministry of the Coal Industry in Moscow, headed by Mikhail Shchadov (who had replaced Bratchenko in December 1985). In other words, the "middle man" in the form of the Ukrainian ministry had been eliminated in favor of a two-tier system. This change could be interpreted as a logical process of perestroika, but it may also have been a form of chastisement of the former ministry for its manifestations of independence, demonstrated in a virtual running battle with the Moscow ministry. Shchadov himself, who was to become a very controversial figure during the coal strike, was present at the inauguration of the two production associations, but had mixed views on the future of the coalfield. He had stated in an interview that the future rise in coal output would be attained predominantly through opencast mines, for example.[14]

On the other hand, he had stated that underground output should be "stabilized" at the present levels and would eventually rise because of the construction of large new mines and as a result of reequipping the existing mines with improved technology. There were three main reasons why the Donetsk Basin should be "saved" as a coalfield, according to Shchadov. First, the coalfield accounted for almost all the valuable anthracite produced in the Soviet Union. Second, it remained the main supplier of high quality coking coal for the metallurgical enterprises of the European part of the country. And, finally, there was a sentimental attachment to the region, with its revolutionary and military traditions, and its history as the home of the Stakhanovite movement.[15] A significant omission from his statement was the needs and futures of the miners themselves, but in itself this was hardly unusual and indicated that the tradition of self-sacrifice on the part of the miners was to continue, at least in the view of the Soviet Coal Minister. The "old thinking" was not yet dead.

In Shchadov's view, there were some major problems ahead, and they lay once again partly in the failure to develop adequate machinery. Materials, he declared, were being produced at a low technical level. Scientists were not fulfilling their role in the pro-

---

14. *Sovetskii shakhter*, No. 11, 1987, p. 6.
15. *Ibid.*

cess of transferring research products into industry. Of the scientific cadres in the industry, he noted, only four out of 150 with a doctor of sciences degree were under fifty years of age, and most had exhibited a marked reluctance to carry out their work. Some institutes were now being abolished for failing to undertake any assignments. The minister was thus faced with a problem of carrying out reconstruction that envisaged more advanced machinery while the research institutes entrusted with developing new technology were in a parlous state.

Yurii Baranov, a manager working at the Donetsk coal basin, was interviewed in 1988 by the Kiev weekly newspaper for Ukrainians abroad, *News From Ukraine*. He expressed the view that the rate of restructuring in the coalfield was well behind the proposed levels. He noted that despite the new Law on Labor Collectives, some legal problems had arisen:

> The state of things here is far from satisfactory. For instance, a collective decides to sack a worker due to his violation of safety regulations. Labor teams agree, mine organizations approve the decision, but somebody in the higher echelons will cancel it just in case. The labor collective council has no legal rights.[16]

Baranov said that because of the poor supply system ingrained in the process, shady deals and recourse to the black market were required in order to obtain the necessary equipment. Self-financing now existed on paper, he stated, but in reality would not be introduced until January 1, 1989. Thus 1988 was to be the preparatory year and not the pioneering year for self-financing (the situation was no clearer, incidentally, in the summer of 1990). Baranov also commented that, as a manager, he was now supposed to make independent decisions and "denounce everything that is obsolete." He was being hampered, however, in this task by the ever increasing plan targets. He was compelled to fulfill the plan at any cost, even if it turned out to be unprofitable. The entire interview was a manifestation of bitterness and disillusionment with the prospect of any reform of the Donetsk coal basin.

16. *News From Ukraine*, No. 2, January 1988.

In 1987, Surgai participated in a further debate within the Soviet coal industry that took place around a new document entitled "The Kuznetsk Variant," which was based on research conducted by several academicians from the USSR Academy of Sciences, under the leadership of M. Styrikovich. The essence of this document was that it was no longer feasible for the Soviet authorities to invest heavily in Donbass coal, because it had become too difficult and expensive to mine. Instead, it was proposed to raise substantially the coal output from the Kuznetsk coalfield in Siberia, from 144 to 300–335 million metric tons annually by the year 2000. Surgai responded to the document in the newspaper *Sotsialisticheskaya industriya* in May 1987, and then an abridged version of this same interview appeared later in this same year in the weekly *Moskovskie novosti,* indicating that the discussion was still in progress.[17] The comments by Surgai revealed that he was prepared to carry the mantle of his predecessor Hrynko, in defending the future of the Donbass coalfield.

According to Surgai, the contents of the documents indicated that capital investment into the Kuznetsk coalfield was to be boosted by six billion rubles annually, while concomitantly that into the Donbass coalfield was to be reduced by two billion rubles. The Siberian fuel was then to be transported into Ukraine to replace Donbass coal in Ukrainian industries. Such an idea was hardly new, and in fact represented the status quo in that some Siberian coal was already servicing Ukrainian enterprises. The document did, however, represent a serious threat to the future of mining in the Donetsk Basin. If its recommendations were accepted, then thousands of Ukrainian miners would be looking for new jobs. Surgai maintained that the "Kuznetsk Variant" would preclude the simultaneous development of two major coalfields in the USSR, thereby posing a threat to the entire energy program of the country. He pointed out that the claim that the Donbass had exhausted its resources was at best a half-truth. Although the best seams had been depleted, new seams had been developed with reserves of some nine billion tons, one-third of which consisted of valuable coking coal.

To this statement, the correspondent responded scornfully that the minister (for the first article appeared before the abolition of

---

17. *Sotsialisticheskaya industriya,* May 28, 1987; *Moskovskie novosti,* supplement to issue 34, August 23, 1987.

the Ukrainian Coal Ministry) was not dealing with realities. While new coal reserves were being brought into exploitation in the Donbass, coal output continued to decline. Extraction had been "stopped or reduced" at seventy-eight mines since 1982, and this trend was continuing. At this point in the debate in the earlier version of the interview, Surgai cited examples of mismanagement at the all-Union level as the main reason for the declining Donbass output. Plans were made to reconstruct several old mines in the Voroshilovhrad Coal Association in 1971 and were sent to Moscow. Nine years later they remained in the hands of the USSR Ministry of the Coal Industry, untouched and unresolved. As a result, the reserves at the mine in question were depleted, and new plans had to be drawn up hastily that turned out to be inadequate. In short, one sees again the now familiar argument of centralization leading to the decay of Ukrainian industry but, interestingly, from a member of the bureaucratic hierarchy, and not from the people.

Other arguments were also advanced by both sides. The correspondent for *Sotialisticheskaya industriya,* evidently a supporter of the Kuznetsk Variant, made reference to the danger of sudden methane gas explosions in the Donbass mines (see below) and the fact that over two-thirds of cleared faces had unstable and false roofs. Today, he stated, the cost price for a ton of fuel in the Donetsk Basin was twice that of the Kuznetsk Basin, while labor productivity was half that of the latter. Consequently, it was advantageous, in the correspondent's view, to bring coal from Siberia to the European USSR rather than mine it in the Donbass. Surgai's retort was to attack the transportation system's inefficiency and to point out the lack of a suitable infrastructure for workers in Siberia. He claimed that about 600,000 tons of coal were being lost annually in transportation and that if the energy costs incurred in transportation were added to this figure, then the total financial loss would be five times higher.

Moreover, railway transportation in the country was already heavily overloaded. The calculations made by the Ukrainian Coal Ministry, in contrast to those of the authors of the Kuznetsk Variant, revealed that the proposed transfer of coal development from one region to another would result not in a saving in capital costs of four billion rubles (as the report had stated), but in a loss of at least two billion rubles. In Surgai's opinion, the Variant's implementation would also lead to the unemployment of 350,000 Donetsk Basin miners. Where would they be employed, he asked? He argued that the expenses incurred in reemploying these people, combined with the cost of moving 110,000– 140,000 workers to the Siberian coalfield would amount to 6.5 billion rubles.

In June 1987, Surgai's statements received support from a doctor of geological sciences, V. Bilokin, who decided to advance several cogent ideas in a series of articles in a Ukrainian newspaper to dispel what he referred to as "the myths created by the Ministry of the Coal Industry of the USSR." He noted that because of the losses incurred in transporting coal to the Donbass, the real cost price of Siberian fuel was almost double that stated officially. He also disputed the claim that because Donbass coal was contained predominantly in thin seams, its quality had deteriorated. On the contrary, he stated, "coal from thin seams is 1.5 to two times better quality than coal extracted from thick seams." Furthermore, he wrote, the Donbass coal had been extracted in an irrational manner: the easily accessible reserves had been used up first, after which the mines had been shut down and the remaining reserves "written off."[18]

One example cited by Bilokin was that of the coal mine "Lyutkova" in the Krasnodon Coal Association. Brought into exploitation in 1964, it had enough reserves of coking coal to last more than sixty years. Forty percent of the fuel was concentrated in thin seams. On this occasion the Ministry of the Coal Industry of the Ukrainian SSR decided to close down the mine on the grounds that its neighbors were now being exploited. The mine was duly shut down in 1986, after the miners had extracted only 38 percent of its prospected seams; the unrecouped capital investment amounted to twenty million rubles. Similar cases were in evidence throughout the East Ukrainian coal mines. In theory, vast reserves remained. While Bilokin's example was one in which the Ukrainian ministry was criticized, the habitual complaint had been that the authorities in Moscow—which was after all the source of all capital investment and all planning

18. *Robitnycha hazeta*, June 16, 1987.

decisions—were prepared to see the massive Ukrainian coalfield sink into an irrevocable decline.

The lack of investment in the 1980s into the Donbass had begun to have repercussions on the industry. In 1989, Surgai noted that sixty-three mines in the Donetsk Oblast (41 percent) had been exploited for more than forty-five years, thirty-three (22 percent) for more than seventy years, and three for over 100 years but that 75 percent of all mines had not been refurbished for over twenty years. The basic reason for this, he declared, was insufficient capital investment in Ukraine's major coalfield: Donbass mines could not be expected to meet production targets without improved equipment. Yet over the previous seven years, the cost price of such equipment had doubled without any corresponding rise in its productivity. In addition, a large quantity of promised equipment, including 206 coal wagons and twenty "mechanized complexes," was never delivered to the coal mines.[19]

The growing ecological awareness in Ukraine was also having its repercussions on the coal industry because the mine authorities were now obliged to expend valuable funds on environmental safety that reduced any meager profit margin that existed. In particular, the ecological lobby had been complaining bitterly about the problems caused by the coal industry in the western Donbass, where coal workings had caused the land to subside, imperiling agricultural production. The oblast party committee in Voroshilovhrad (Luhanske) revealed early in 1989 that coal enterprises in the oblast had to date spent more than seventeen million rubles on nature protection measures, of which 6.4 million had been used for waterprotection schemes, 4.7 million on cleaning water polluted by coal mining, and 1.3 million on restoration of spoiled land.[20]

WORKING CONDITIONS BEFORE THE STRIKE

Although the hazards of working in underground coal mines are well known, before the Gorbachev period little public attention was given to coal mining accidents, other than official condo-

19. *Ugol' Ukrainy,* May 1989, pp. 1–2.
20. *Radyans'ka Ukraina,* February 15, 1989. See also *Robitnycha hazeta,* May 18, 1990.

lences to relatives of the victims in the press. With glasnost and after Chernobyl, as with many other facets of Soviet life, the situation changed. If there was a serious accident, then a government commission was to be established to determine its causes, with a full and detailed inquiry, the results of which would be published. This change in direction highlighted the difficult life of the coal miner as never before and soon there were various scientific inquiries as to the state of health of coal miners, which areas of their lives gave them cause for concern, how they wished to change the situation at the coal mine. Before long, the miner recognized that despite various advantages of coal mining, such as relatively long vacations and higher pay for face work, his work conditions were appalling.

One can provide a brief review of some typical accidents, over the decade 1979–1989, in the Donetsk Basin. In August 1979, a severe accident occurred at the Molodohvardeiska mine of the Krasnodon Coal Association. The official report stated that a methane gas explosion had caused deaths and that the authorities were taking steps "to eliminate the consequences of the accident."[21] Several months later, an almost identical explosion occurred at the May the First Coal Association in Voroshilovhrad Oblast. The Soviet authorities expressed their condolences to the families of the deceased miners.[22] Early in 1981, the Ukrainian Academy of Sciences noted that the number of dangerous mine explosions over the previous decade had tripled.[23] The main reason for such an unfortunate development was that as mines became deeper the chances of an explosion of methane gas had become correspondingly higher. More recently two serious accidents occurred in the Donbass on December 24, 1986, and May 16, 1987, involving loss of life as the result of explosions of methane gas. The first was at the "Yasinovskaya-Glubokaya" mine in Makiivka; the second at the Chaikino mine near Donetsk. The latter accident reportedly could have been avoided with better ventilation in the mine.[24] But there were other, more basic types of accidents.

21. Radio Kiev, August 11, 1979.
22. *Pravda*, April 28, 1980.
23. *Radyans'ka Ukraina*, January 6, 1981.
24. *Sotsialisticheskaya industriya*, December 27, 1986; *Pravda*, May 20, 1987; *Trud*, May 24, 1987. The Russian names of the mines are being used rather than the Ukrainian since only that form was in general usage at the mines

One that was reported more fully provides a notable illumination of the rather disdainful attitude of the authorities toward the coal miners. It concerned the collapse of a mine roof at the "Krasnoarmeiskaya Zapadnaya" (in Ukrainian, *Chervonoarmiiska Zakhidna*) No. 1 coal mine in the Donetsk Basin. The accident took place at the end of a shift, and had left fifteen miners trapped 700 feet underground for thirty-two hours. The miners shared what water they possessed and were able to keep in touch with their colleagues on the surface because the cave-in had not damaged the telephone cable. At the same time, there was great danger of a gas explosion. The mine in question is located in a sandstone region, and the gas abundance of sandstone raises the concentration of explosive methane. Rescue workers were evidently hindered in their efforts by a twenty-five-meter coal seam that ran lengthways across their paths. They used "miners' picks" and literally hewed away at the rock. Shaft sinkers from the "Krasnoarmeiskaya" mine were assisted by miners from the neighboring "Stakhanov" mine, and coal hewers arrived from Horlivka during the latter stages of the rescue operation.[25]

The main point about this particular accident was the way in which it was reported by the Soviet press. Previous accidents, such as the similar roof collapse at the "Torez" mine in the same coalfield in October 1984, were reported in relatively low circulation newspapers. The "Torez" accident was reported in the Kiev workers' newspaper *Robitnycha hazeta* which, together with its Russian-language counterpart, *Rabochaya gazeta*, had an estimated circulation of only 100,000 in 1975. Yet the "Krasnoarmeiskaya" mine accident was written up in the far more popular *Sovetskii sport*, a newspaper with a circulation of around four million copies. The question arises why, after years of virtual secrecy about the details of coal mining accidents, which only served to illustrate the dangers of working in the Donbass mines, the authorities chose to disseminate news of this event in a mass circulation newspaper.

The answer lay in the way the accident was reported. Essentially it was trivialized in order to make it entertaining to the

themselves.
25. *Sovetskii sport*, March 17, 1985; *Trud*, May 17, 1985. In neither of these reports was it made clear when the accident actually took place.

general reader. The reasons for the accident were dealt with in a few sentences (it was caused by a geological protuberance, or "lens" formation in the sandstone that made contact with the surrounding rock). Instead, the emphasis was placed on the character and resilience of the men trapped underground. The accident was "a test of the miner's courage," and the survival of the fifteen men under such conditions was attributed to their "excellent physical hardening, will and endurance." The trapped miners were heroes fighting against nature's elements, but because they were miners, they might have been expected to survive. Naturally, the reader might conclude, the survivors of the Stakhanov inheritance might be expected to react in this heroic manner. It was less likely that the reader would ask the most pertinent question, namely, how widespread were such accidents and were they primarily a result of the worsening geological conditions in the country's oldest coalfield.

The authorities in the past had tended to blame the careless manager or even worker in cases of accidents. Several years ago, the Soviet trade union newspaper, *Trud*, reported that mining accidents were to some extent the fault of miners who did not follow safety regulations. Far more to blame, however, in the view of the newspaper, were those engineers and managers who ensured that safety rules were adhered to "only when it is known that the inspector from the State Technical Inspection Bureau *(Gostekhnadzor)* will be visiting." The violations were said to be so serious that inspectors often had to suspend all work in mines until the danger had been eliminated. In 1979 alone, 154 people were demoted or dismissed on the recommendation of the Inspection Bureau, while thirty-seven people were placed under investigation.[26] Two years later, Mikhail Srebny, Chairman of the Union of Soviet Coal Miners, interviewed by Radio Moscow, declared that only 5 percent of all mining accidents were caused by faulty equipment. In other instances, he maintained, the carelessness of the miners had led to the mishaps. The statement appeared to contradict the report in *Trud* noted above. But it did raise the question of what safety mechanisms had been introduced into the mines.

Such measures fell into two categories: personal health and mine safety. In the former area, the USSR had claimed that min-

26. *Trud*, June 18, 1980.

ers enjoyed more than adequate protection. According to one observer, the Moscow Mining Research Institute had demonstrated that the incidence of silicosis (coal dust inhalation) among Soviet miners was several times lower than in other countries. The maximum limit had been set at ten milligrams of coal dust per cubic meter of air. Each Soviet mine, he stated, possessed a twenty-four hour medical service, and every year about 200,000 miners underwent preventive treatment at one of 200 outpatient clinics. A specialized sanatorium was said to be under construction in Rostov Oblast in the easternmost sector of the Donetsk Basin coalfield.[27] Methane gas sensors had been installed at coal faces by early 1982. These reportedly controlled the electricity supply and switched off power when the permissible level of gas concentration had been exceeded. Coal cutting combines were equipped with dust filters and extra ventilation was helping to keep the dust in check. Yet with the deepening of the mines, the costs of installing such equipment had risen and it is far from clear whether every mine possessed such safety installations in the 1980s.

In 1984, the Ukrainian Ministry of the Coal Industry carried out a survey of over 1,600 young miners to monitor their dissatisfaction with various aspects of work in the Donetsk mines. First, the miners were asked to give their opinion of conditions and social services, such as "relations in the collective," public transport to the workplace, and "interest in work." They were then asked to estimate the importance of each separate category. In this way, it was felt that their main concerns could be adequately ascertained.[28] The survey was important because it indicated not so much that the miners were unhappy (the whole point of the survey was to air grievances) but where the priorities of the miners lay.

About 19 percent of those surveyed expressed unhappiness with the sanitary and hygienic conditions, which was evidently a concern of a republican conference held around this same time in Donetsk. Yet only 2.5 percent out of the 19 percent considered this factor to be of major significance. By contrast, almost 37

27. Radio Moscow, December 21, 1981.
28. *Sotsiologicheskie issledovaniya*, No. 2, 1984, pp. 14–15.

percent of interviewees were dissatisfied with transportation, and about 75 percent of those who expressed their discontent felt this to be a matter of importance. A similar result was found in the sphere of recreational facilities. Curiously, the survey did not reveal much discontent with the working conditions at the mines themselves, and 97.1 percent expressed satisfaction with the situation in the work collective. One wonders therefore whether the young miners surveyed may have been guarded in their responses, either through habit, or because they feared for their job security otherwise.

Around this time, at the very beginning of the Gorbachev period, Soviet writers liked to demonstrate that Ukrainian miners were generally happy with their present existence. During the British coal miners' strike of 1984–85, TASS reported that Ukrainian miners had expressed great sympathy toward their British counterparts. Consequently, after meetings at fifteen mines in the Donetsk Basin, attended by over 100,000 miners, about 500,000 rubles had been contributed to a relief fund for the British strikers.[29] According to one of the contributors, however, there was nothing voluntary about such contributions. Rather the authorities had simply deducted ten rubles from each miner's monthly wage packet, whether the miners approved or not. As this particular miner put it, "Ten rubles is a lot for a man with a family to support."[30] The image of the affluent Donbass miner aiding his suffering British comrade was thus hardly an accurate one.

The Soviet authorities had become particularly sensitive on the issue of miners' welfare and working conditions. In January 1985, the London *Times* devoted a lead editorial to the topic "The Life of the Soviet Miner," which spoke of "miserably low wages" and an increasing number of mining accidents and illnesses, as reported in *Trud*.[31] One week afterward, the latter newspaper responded to these "blatant lies or boundless stupidity," maintaining that the take-home wages of the Soviet miners were well above those of their British counterparts and that the "injury rate" (it was not clear what was meant by this term) in

29. TASS, December 20, 1984.
30. *Radio Free Europe/Radio Liberty Soviet Area Nationality Survey*, December 1984.
31. *The Times*, January 8, 1985.

Soviet coal mines had fallen by about 13 percent between 1980 and 1984.[32] *Trud* also encouraged Soviet miners to respond to the *Times* editorial, and several miners evidently did so because their responses were published by the Soviet newspaper in its edition of February 23, 1985. However, evidence suggests that the *Times* editorial had exposed a raw nerve by providing one of the rare accurate accounts of the situation in Soviet underground mines at this time.

One of the most revealing examinations of the Donbass miner was provided in a series of articles by two special correspondents of the newspaper *Sotsialisticheskaya industriya,* V. Andriyanov and G. Dorofeev, in the summer of 1988. The series was entitled "The Life of a Miner," but the authors confined their investigations to the Donbass mines. Their findings constituted a damning indictment of the way the industry was being run in the area and indicated that there had been an almost total disregard for the health and living conditions of the miners. After the final article in the series had appeared, the Collegium of the USSR Ministry of the Coal Industry and the Presidium of the Central Committee of the Trade Union of Coal Industry workers held a meeting to discuss the shortcomings raised by the articles. At the meeting, which was chaired by the Coal Minister, Mikhail Shchadov, there was general agreement that the authors' criticism was justified: Nikolai Surgai, for example, conceded that three million man-days had been lost in 1987 in Donetsk Oblast because of illness; V. Poltavets acknowledged that the situation was much the same in Voroshilovhrad Oblast.[33]

The official acknowledgement of the problems detailed in the articles had come very belatedly, however. As the authors made clear, they were portraying problems that had been around for more than a decade. They noted that the miners had a relatively short lifespan and that labor conditions had had a direct influence on the incidence of mortality. They spoke of a significant rise in high blood pressure, heart disease, and mental disorders among miners, and stated that the incidence of sudden coronary death had doubled over the previous ten years. They discovered, moreover, that the percentage of those with heart disease who

32. *Trud,* January 16, 1985.
33. *Sotsialisticheskaya industriya,* August 24, 1988.

were working in mines where the temperature was over twenty-six degrees celsius (eighty-nine degrees Fahrenheit) was almost double that of those working in shallower mines.[34]

According to the authors, the USSR Ministry of the Coal Industry had been so indifferent to this state of affairs that it had repeatedly ignored letters on the topic from local experts. Professor V. Sukhanov, head of a department of labor hygiene, was quoted as saying: "Essentially the Ministry of the Coal Industry and the mine managements are not concerned with the health of the miner." Letters from the Donetsk Research Institute of Labor Hygiene and Occupational Diseases to Coal Minister Shchadov requesting information about measures being taken to combat the high rate of illness among Ukrainian coal miners had received no response.[35] The other articles in the series, while less dramatic in tone than the first, also focused on important aspects of the coal industry in Ukraine: on the cut in the rate of housing construction for the miners since 1987; on the severe ecological damage caused to the environment by mining; and on the very high rate of manual labor in Ukrainian coal mines, in which 50 percent of the work below the surface was still being carried out by hand.[36]

Ironically, the appearance of these articles coincided (more or less) with the publication of the Director-General's report to the Twelfth Session of the Coal Mines Committee of the International Labor Organization (ILO), which relies on information requested from its member states. According to this report, as cited by a Ukrainian monthly journal, between 1980 and 1985, fatal accidents in Ukrainian coal mines declined by 11.6 percent, and from 1976 to 1986, the incidence of pneumoconiosis fell by 80 percent as a result of the development of preventive measures. In the journal's introduction to the excerpts from this report, it was even suggested that the experience of the Ukrainian SSR in dealing with problems in the industry would be useful as a pointer for other coal-producing countries.[37] It does not appear that the report in question was very accurate and it provided a very misleading picture of the true state of affairs in the coal mines.

34. *Sotsialisticheskaya industriya*, July 26, 1988.
35. *Ibid.*
36. *Sotsialisticheskaya industriya*, July 31, 1988; and August 2, 1988.
37. *Ukraine*, No. 7, 1988, pp. 28–29.

By mid-1988, further revelations were being made evident. A significant article on the situation was published by M. Kabanets, Chairman of the Coal Industry Sector in the Department of Heavy Industry within the Central Committee of the Communist Party of Ukraine. He stated that 75 percent of the Donbass mines had not been reconstructed for twenty years. In about one-third of them, he went on to say, the depth of exploitation was between 800 and 1,000 meters. (Even the ILO Report, as quoted in the Ukrainian journal, referred obliquely to the danger of work in deep mines; it reported that "the number of coalminers working deep seams at temperatures exceeding those specified in the safety regulations will soon be reduced by 13,000.") Kabanets also mentioned that about 17 percent of the seams currently being exploited were subject to sudden emissions of gas.[38]

Kabanets projected three ways in which the industry should develop in the immediate future: 1) a restructuring of administration by means of a considerable reduction of the governmental apparatus and an increase in the "independence" of work collectives; 2) a transition from administrative-command to predominantly economic methods of management and a move to full self-accounting and self-financing; and 3) a democratization of the work process by application of the rules of self-management to the collectives. Some of these measures were in the process of being initiated. The abolition of the Ukrainian Coal Ministry, for example, had reduced administrative personnel in the branch by 48 percent and the monthly costs by 25 percent. Kabanets commented accurately, however, that "restructuring" of the former ministry had hardly begun.

Another attempt to improve the situation in the Ukrainian mines was the gradual reduction of the number of those employed in manual labor by 8,000 to 10,000 workers annually. In turn, such a maneuver could hardly be expected to receive widespread report among those miners being deprived of jobs. If perestroika signified unemployment, then the coal miner was not in favor of perestroika. But why was his situation so difficult? In theory, the life of the Donbass miner was relatively comfortable in the 1980s. His working week was only thirty hours, compared to thirty-six to thirty-eight hours in other coal mining regions. In

38. *Pid praporom leninizmu*, No. 14, 1988, pp. 89–92.

1980, on the advice of the Institute of Work Hygiene and Professional Health Care, a six-hour workday had been accepted for all Soviet miners, with a 20 percent norm reduction in deep mines with excessively high temperatures. Work shifts were reduced by 10 percent for every two "above norm" degrees of temperature. Retirement was permitted at the age of fifty, and by 1979, the miners' pension, which was up to 170 rubles per month depending on the length of time worked in the industry, was higher than most pensions in Ukraine a decade later.[39]

Turning to wages, in September 1981, the Central Committee of the CPSU and the USSR Council of Ministers had issued a decree "On Raising the Wage Rates and Salaries and Improving the Organization of Pay for Workers and Employees in the Coal Industry and Mine Construction." Workers engaged in underground operations received an average wage increase of 27 percent; those involved in surface operations, 23–24 percent; and those in other work, 18–20 percent. The increased rates of pay were introduced into the Donbass region in the first quarter of 1982.[40] After the wage rise, it was said that some underground miners in the Donbass were earning between 350 and 400 rubles a month, which was almost double the average monthly wage in Soviet industry. But even this pay scale was to become intolerable for the coal miner in the economic conditions that were in place in 1989—it remained in place until after the strike. Moreover, it would be very misleading to judge the miners' real situation simply from the figures provided.

For one thing, the wage rises were linked partly to productivity, and as plan targets were missed, as was usually the case in this coalfield, then the pay packet declined accordingly. Saturday work on a voluntary basis had become the norm. Thus when a "subbotnik" (voluntary working Saturday) was announced at the Twenty-Seventh Party Congress in February 1986, one delegate, Petro Venger, remarked that in the Donbass coalfield "Saturday is a normal working day." Sunday work had also become the norm. In 1982, the coal miners' union acceded to management's demands to work two Sundays in the month. Yet 150 managers were said to be violating labor codes by introducing Sunday

39. *TASS*, August 6, 1979; *Visti z Ukrainy*, No. 38, September 1980.
40. *Pravda*, September 13, 1981; *TASS*, February 10, 1982.

work throughout the month, with only standard rather than double pay for overtime.[41] Even earlier, in 1980, it had been reported that in the Donbass coalfield miners were "constantly being deprived of their Sundays off."[42]

This same problem was also broached by V.S. Shatalov, a deputy chairman of the State Mining Safety Inspectorate, following Mikhail Gorbachev's visit to Ukraine in February 1989. Shatalov maintained that safety regulations were often being breached in the coalfield and that such violations were the chief cause of accidents. Among them, he singled out the fact that there was no such thing as a "day off" in the Soviet coal industry. Coal miners in Czechoslovakia, Poland and Hungary had both a day off during the week and their Sundays free, he pointed out. On these rest days, basic repair and maintenance work could be carried out. In Soviet coal mines, on the other hand, such work was neglected in the interests of raising or maintaining output levels, and as a result more accidents were occurring. Shatalov revealed that mining equipment was being inspected only after an accident had taken place.[43]

When irregularities took place, those responsible—chief engineers or heads of mine sections—could be penalized or fined, but such punishments had become so routine that they had ceased to have any deterrent effect. Shatalov cited the cases of two officials of the Donetsk Production Association, one of whom had received six admonishments in the course of a year, and the other twelve. In his view, work safety had not been a main priority of coal enterprises, partly because the costs to the state of major accidents did not affect the financial position of the enterprises. He maintained that occupational injuries must be reflected in the enterprises' balance sheets if they were to make any impression on the leaders of those enterprises.

There were also technological problems in the coal mines, according to Shatalov. In particular, adequate technology to operate deep mines with thin and sloping seams had clearly not been developed. He made reference to "gross engineering miscalculations," the failure to resolve the problem of gas seepage in Don-

41. Radio Moscow, February 13–15, 1982 (series of reports).
42. *Trud*, September 21, 1980.
43. *Ekonomicheskaya gazeta*, No. 7, 1989, pp. 17–18.

bass mines, and poor ventilation in the mines also of the Kuzbass Basin in Siberia. He revealed that more than 600 miners were dying each year in accidents caused mainly by tock falls during the extraction of coal, the retimbering of mines, and the transportation of coal and miners underground. A key and related problem, in his view, was the failure of research institutes to provide the necessary technology to deal with the worsening mining conditions. Some 45 percent of underground technology was said to be unreliable, as was 29 percent of ventilating and 70 percent of face clearing machinery. Great claims had been made about scientific progress in the branch, but inventions displayed with great pomp and flourish in exhibitions had proved useless in service. Even plans to equip mines with loudspeaker systems—especially important at the time of an accident—had not been carried out.

The miner, then, faced an uncertain and dangerous existence. Many stories surfaced about the poor health of individual miners, placing names on the anonymous pictures of the victims of silicosis and other illnesses. It was also apparent that it was becoming increasingly difficult for the mining family to survive economically. In the summer of 1989 an article was published that illustrated these two phenomena. It related the story of a man who went to work at a Donbass coal mine in 1956, at the age of eighteen. Within seven years he had silicosis of the lungs and was designated as an invalid Class III. However, the mines were suffering from a labor shortage so that he continued to work underground. He was permitted a pension of seventy rubles and could earn an addition 140 at the mine without penalty. But continual wage and price rises quickly reduced the living standards of his family: "my children began to ask for goods that I could not afford."[11]

Consequently this miner was obliged to return to face work, but this led to a decline in his health and categorization as an invalid Class II. He was thus forced to retire from the mine altogether and take a job as a carpenter. Yet even then his position was far from simple. The size of the pension was based on the income he was receiving in 1962 just before he became an invalid and if his wage rose above the 140 rubles, he was obliged to pay a certain amount back to the mine for being "overpaid." Consequently, even while employed as a carpenter (hardly the

44. *Sotsialisticheskaya industriya,* May 7, 1989.

highest paid job in the country), he paid thirteen rubles a month
to the coal mine, the source of his illness in the first place and
his work location for thirty years. At the same time pensions
were said to be so meager that most of the miner's peers were
also continuing to work in order "to stay alive."

THE COAL MINERS' STRIKE

It is well known that the coal miners' strike of July 1989 began
outside the Donbass, and while we have argued that the situation
for the miners in the huge East Ukrainian coalfield was deplorable, this does not necessarily indicate that conditions were better
elsewhere. The strike began as a grassroots movement and grew
into the most serious test of the Gorbachev administration to
date. But the initial spark—rather than the more fundamental
reasons behind the strike—seems to have been the insensitivity of
local and party officials to the miners' demands. According to
Boris Yeltsin, this insensitivity, in turn, was a result of political
reforms. Now that these regional "bosses" knew that they would
no longer be elected automatically to their old posts, they had
become "inactive" and simply could not care less what the miners wanted.[45] In the town of Bryanka in the Donetsk Basin, a
strike began because the equipment in the mine failed and the
administration did nothing for four days. Shortly thereafter the
miners had established strike committees that began to take over
the functions of the city government.[46]

In the Pavlohrad Coal Association in Dnipropetrovsk, it was
also reported that local Soviets and party organizations were
"deaf" to the needs of the miners, compelling them to take the
extreme step of a strike. A. Mironenko, the First Party Secretary
of Pavlohrad region, arrived at the headquarters of the Association on July 18 to determine the reaction of the miners to the
strikes that had broken out in the Kuzbass. Not only were the
miners of Pavlohrad in sympathy with their fellows in the east,
but they were on the verge of striking themselves and by 5pm on
this same day, the first miners had taken to the streets. By the
following day, all eleven mines of the association were on strike.

45. Cited by Andrew Katell, *Associated Press*, July 24, 1989.
46. *Literaturnaya Rossiya*, September 29, 1989.

Miners complained of the "dictatorial methods of the administration," of the environmental damage being caused by the coal mines in the region, and of the poor housing that had been reserved for them. These complaints had evidently surfaced a year earlier, but there had been no reaction to them either from the local or national leaders, or from the USSR Ministry of the Coal Industry.[47]

While the miners may have reached the end of their patience, it also appeared that the strike was exceptionally orderly. Liquor stores were closed, there was very little or no civil disorder, and the crime rate in many cities went down considerably during the strike. It was as though, after years of incompetent leadership by the local authorities, the miners simply swept them aside and took over. Not infrequently, the strike committees soon uncovered a variety of corrupt activities. In one state store in Bryanka, for example, champagne and other goods were found to be missing from the shelves because the store owner had a list of "special clients" for whom they were reserved. While the city of Bryanka was in debt to the tune of 600,000 rubles this one store owner was said to have accumulated goods in her home that were worth 60,000 rubles.[48] Such instances were hardly unusual and had long been a complaint in cities in the Soviet Union. But it seemed to the strikers that it was only when their committees had been formed that action was taken on such matters.

There was also consensus that the origins of the strike lay many years ago in Soviet history. One spokesperson argued, in a round-table discussion, that the most important date to recall was 1928, when M.P. Tomsky, the head of the trade union organization was arrested by Stalin (he later committed suicide), after which the interests of the workers were neglected.[49] Another cited the Khrushchev period, when the military had crushed an attempted strike of coal miners. If this had not happened, he stated, then economic reforms might have begun "twenty years earlier."[50] One journalist raised the poignant question: "who is it

---

47. *Ekonomicheskaya gazeta*, No. 32, August 1989.
48. *Literaturnaya Rossiya*, September 29, 1989; *Stroitel'naya gazeta*, July 19, 1989.
49. *Literaturnaya Rossiya*, September 22, 1989.
50. *Sotsialisticheskaya industriya*, July 28, 1989. The speaker's arithmetic was not very accurate because twenty years earlier, Brezhnev had been in office rather than Khrushchev. But the point is clear enough.

that is protesting?" Why, he responded, it was the working class, "the same class that for years had been portrayed in the press as happy and smiling, and full of optimism." He maintained that the slogan that the workers ran the factories had now been shown to be a false one. In his view, the miners' strike was the result not merely of the neglect of popular needs, but of this "lie" as well.[51]

The local authorities reacted to the strike in hamfisted fashion, with truculent demands that the miners should return to work. Their position on the face of it seemed powerful enough. This was the first major strike in Soviet history. What was later to become an almost routine "downing of tools" was at this time shocking to many. The miners surely could not win this battle with the state. Thus the authorities resorted to familiar tactics. One appeal, for example, was initiated from the Mariupil metallurgical combine workers to their "brother miners." If fuel was not forthcoming from the mines, the combine would not be able to produce quality material. In addition, the steel workers had become concerned about the safety situation as a result of the lack of fuel supply (the logic of this statement is somewhat difficult to decipher). They were anxious also about the loss of production and how this would affect the "lives of the many thousands" who worked there. The steel produced by the combine was exchanged at the docks for foreign goods such as medical supplies, razor blades, washing powder, tea and coffee.[52] The implications of such an article were clear enough and amounted to emotional blackmail.

Another account spoke of the "irresponsibility of the strike committees." While the government representative was meeting with representatives from Donetsk, the miners were sitting in the city's main square, ignorant of the fact that "essential supplies" for power stations and industrial factories were being cut off. Those that did want to return to work, it was reported, were regarded as "blacklegs" and "strike breakers." Even though the miners had suffered terribly from the conditions of the mines, they were now being blamed by the authorities for leaving their mines because lack of maintenance might cause serious longterm

51. *Izvestiya,* July 22, 1989.
52. *Pravda Ukrainy,* October 24, 1989.

damage to the mines, such as underground fires. The loss of production by those miners sitting in the square, the article noted, amounted to six to seven million rubles each day, and this would mean that there was less money to pay doctors, teachers and other professional workers. Yet, the author wrote, even though Gorbachev had appeared on television and convinced many miners to return to work, this had had no impact in Donetsk, and very little in Voroshilovhrad.[53]

The question that arises is: why should such considerations have made a difference to the miner? A strike may ultimately be regarded as a selfish act, but in this case it was a step taken when all other avenues had been closed off. In Ukraine, the strike took on an especially militant form in the small Lviv-Volyn coalfield, with huge demonstrations of striking miners in Chervonohrad. There, the Lviv Oblast party leader, Ya. P. Pohrebnyak, appeared before the strikers at one point to assure them that 78 percent of their demands had been met and that the others could be worked out while the miners were back at work. The speech was to no avail as angry miners, one after another, listed their various grievances and noted the alarming incidence of sickness among them.[54] The strike was coming to an end by the evening of July 19 in Kemerovo coalfield, but it was just reaching its climax in the Donetsk and Lviv basins, in Rostov on the-Don and Vorkuta.

During the strike itself and its aftermath, some new information emerged on the grievances of the miners. A strike leader from the Donetsk region, A.S. Dubovik, noted that in the Donbass, three to four lives were lost for every million tons of coal extracted. In another account from the same area after the strike, it was revealed that forty-four miners had been killed in July 1989, sixty-seven in August and altogether, 431 in the first eight months of 1989.[55] In several instances, there were bitter complaints about the length of the working day, which was being extended by some hours because of the time it took the miners either to reach the mine, or to get to the coal face once work had begun. And virtually every strike committee established through-

53. *Sotsialisticheskaya industriya,* July 25, 1989.
54. *Lenins'ka molod,* July 25, 1989.
55. *Literaturnaya Rossiya,* September 22, 1989; *Sotsialisticheskaya industriya,* September 15, 1989.

out the coal mines of the Soviet Union listed housing problems as one of their paramount concerns.[56]

While each individual coalfield had a separate list of demands with numerous points—the Chervonohrad Coal Association had sixty, for example—we will confine our inquiry to the Donetsk Basin, since this is the main focus of the chapter. The main points, listed as an agreement between the strike committees of this region and the government, were as follows:[57]

1. Mines and coal producing associations in the Donbass are to be given full economic and judicial independence. The industry is to be reorganized to include different types of ownership by the state, cooperatives, leasing and shareholding, etc.
2. Coal producers are to have the right to sell their products both within and outside the country at a price higher than the current official price for coal.
3. Price changes are to be undertaken under the supervision of the State Committee for Prices and the USSR Ministry of the Coal Industry.
4. Within one week, the Ministry of the Coal Industry is to draw up plans to be shown to the USSR Council of Ministers for restructuring of the payscale, and to create conditions in which labor productivity will increase.
5. From August 1, 1989, the local enterprises themselves will determine any changes in wage scales and work norms.
6. From July 1, 1989 [i.e., retroactively], miners will receive an extra 40 percent additional wages for night work and 20 percent for evening work.
7. From August 1, 1989, the transit of miners from shaft to the coal face will be considered part of normal working time.
8. The government promises to investigate illnesses connected with coal mining.
9. Miners whose health had been affected by their work are to be given compensation and/or pensions, the rates of which are to be agreed upon by the strike committees and the mine administration.
10. Any underground worker is entitled to receive a pension after twenty-five years of service, while certain types of work-

56. See, for example, *Pravda Ukrainy,* July 28, 1989.
57. The demands were published in *Vechernyi Donetsk,* September 25, 1989. Altogether there were forty-seven on the original document, among which I

ers may receive a pension after twenty years. If they are over 45, the pension is to be 60 percent of the average wage. Those who have worked underground for twenty-five years will receive 70 percent of the average wage, and those with ten years of service, 50 percent. For men who have worked less than ten years underground and women 7.5, each year of service is to count as two, so that men can retire at sixty and women at fifty-five.

11. There is to be a wage increase for certain groups of workers.

12. The government promises to increase and improve medical services for the miners from September 1, 1989.

13. There is to be an increase in funding for food and other supplies to the mining districts and an improvement in the health and nutriment level of children.

14. There is to be an improvement in the safe operation of heavy machinery.

15. Workers who are transferred from one mine to another are to retain all the benefits accrued at the former mine.

16. Women with children of less than three years of age are to be given leave with no loss of seniority and are to be permitted fully paid maternity leave for eighteen months.

17. There is to be an improvement in the quality of rest homes for miners, especially for those who are sick or injured. From January 1, 1990, sanatoria formerly owned by the Ministry of the Coal Industry will be transferred to the miners' control.

18. Families of miners who are killed at work will receive a year's wages as compensation, and will be provided with a fully equipped apartment within three to six months.

19. No action will be taken against any participants in the strike or members of the strike committees. After the strike is over, the strike committees will work with the unions to resolve any problems.

20. The Donetsk Coal Association is to be liquidated and its building given to orphans and invalids.

21. The state is not to take more than 30 percent of the profits of the working collectives.

22. Sunday is to be a rest day, effective April 1990.

---

consider the following to be the most significant.

23. Underground workers are entitled to forty-two days of holiday each year; surface workers to twenty-four; technical workers thirty, with twelve extra days for overfulfilling the work norm and three for length of service. The maximum permissible vacation is to be fifty-seven days.
24. All miners are to receive an apartment no later than six years after an application for one.
25. Miners are to receive 800 grams of toilet soap per month, effective August 1989. This soap is above and beyond that supplied by the norm.

It appeared therefore that the government had backed down and that the miners' strike had achieved its original goals. Mikhail Gorbachev personally declared that while the route of a strike may not have been the ideal way to express grievances, nonetheless, the miners had solid grounds for taking such action. He identified the miners with the cause of perestroika, and sent one of his main allies in the CC CPSU Politburo, Nikolai Slyunkov, to deal with the strike committees, alongside the Coal Minister, Mikhail Shchadov and Deputy Premier, Lev Ryabev.[58] This was a contrast with Gorbachev's somewhat remote attitude toward the Chernobyl disaster three years earlier when he had given at least covert support to the official secrecy surrounding that accident. Further, as will be seen below, his support of the miners came shortly after the Ukrainian party leadership, under Volodymyr Shcherbytsky, had condemned the miners in no uncertain terms for carrying out the strike.

Both the Communist Party of Ukraine and a group of Ukrainian sociologists analyzed the causes of the miners' strike shortly afterward. At a conference in the Central Committee of the Communist Party of Ukraine held on August 7, 1989, Shcherbytsky noted that the party leaders had to bear much of the responsibility for the strike. It was necessary, he maintained, to undertake a thorough examination of the passive, nonideological position of some party members who had slipped into a position of "nationalism." Vitalii Masol, the Chairman of the Ukrainian Council of Ministers, focused, on the other hand, on the lamentable social and economic situation in many parts of the republic, particularly in enterprises of the coal industry. The recently intro-

58. Soviet Television, July 23, 1990.

duced economic reforms, in his view, were generally not understood by the regional collectives.[59] Borys Kachura, a secretary of the Central Committee of the Communist Party of Ukraine, sharply criticized the actions of party leaders at the oblast level during the strike. Most had adopted a "negative position" and had refused to alter their views despite the course of events, with the result that the party had become isolated from the strikers and lost its credibility. By contrast, he pointed out, informal groups had moved in quickly to exploit the situation. The Democratic Union, the Ukrainian Helsinki Union, and the Rukh had distributed leaflets, appeals and newspapers to the strikers, offering them "political" and financial assistance.

Kachura also analyzed the economic background of the strike, noting that for more than a decade, investment in the Ukrainian coal industry had been geared to raising coal output rather than to the social sphere. By way of illustration, he stated that some 214 million rubles designated for social needs in Donetsk and Voroshilovhrad oblasts in 1986–88, of which thirty-five million alone had been targeted for housing construction, had not been spent. Miners, he went on to say, were not provided with sufficient food products and must either resort to the black market or turn to cooperatives, the high incomes of which had caused great resentment.[60] There had already been signs of an impending strike as early as March and April 1989, he said, when some brigades in the Donetsk Basin had refused to come to work, demanding extra wages for working night shifts. As a result of the strike, Kachura said, there had been a shortfall of 2.5 million metric tons of coal from Ukrainian coalfields. Furthermore, other industries were now beginning to echo the miners' demands. Kachura listed railroad workers, metallurgists, and workers in the construction and chemical industries as among those dissatisfied with the existing situation.

Anatolii Vinnyk, First Secretary of the Donetsk Oblast Party Committee, emphasized the wide gap between words and reality regarding the miners' working conditions. Theoretically, for example, the miners worked only a six-hour day, but it could take

59. *Radyans'ka Ukraina*, August 10, 1989.
60. For an interesting account of the operation of the black market in the Donbass coalfield, see *Pravda Ukrainy*, October 26, 1989.

them up to two hours to reach their place of work. Vinnyk noted that shortages of meat and fruit were less likely to give rise to discontent than the lack of soap and soap powder. The miners, he went on to say, were annoyed by the superficial comments about the situation made by Mikhail Shchadov. In Vinnyk's opinion, the industry was not ready for radical economic reform. A.V. Kasyanov, Chairman of the Voroshilovhrad government, focused on housing. He declared that 83 percent of the miners' residences were under the jurisdiction of the Coal Ministry and that their condition was deplorable. At the start of 1989, he informed, 20 percent of residences lacked a water supply, 26 percent were not connected to a sewage system, 28 percent did not have central heating, 63 percent lacked hot water, and 49 percent were without gas. Like Kachura, he felt that the ministry and its associations were concerned only with coal output and preferred to dictate orders to the mining collectives rather than focus on the miners' social needs.

In Pavlohrad region, described above, the food problem was described as acute. Mykola Zadoya, First Secretary of the oblast party committee, noted that, while the population of the oblast had increased, the area of land under cultivation had dwindled as a result of industrial development. The local population thus required food subsidies, but in the years of perestroika, the volume of such supplies had been halved in the case of meat and reduced by two-thirds as far as milk was concerned. For the first time in many years, the oblast had not fulfilled its plan for building houses, schools and kindergartens, and a portion of the funds earmarked for such purposes—thirty million rubles—had been dispatched as "fraternal aid" to diverse regions of the country. In other words, a "have" region was rapidly becoming a "have not" region.

The focus of the meeting returned time and again to the failure of the party to take initiatives in the miners' strike. Many speakers uttered ritualistic pleas, that the party should take a more active leadership role in the matter and that more radical economic reforms must be introduced. Yurii Elchenko, the ideological secretary, launched a verbal attack on the Rukh which, he maintained, had not disassociated itself from anti-socialist and nationalist groups like the Ukrainian Helsinki Union, the Ukrainian Popular Democratic League, and the Ukrainian Culturological Club. The question is how did such speeches relate to the coal miners? It was an indication, above all, of the falling popularity of the party, especially among the miners. The cause of such a loss was, however, less likely to have been the intrusion of Rukh members into the mining regions than the party's failure to adopt a conciliatory stance toward the miners from the outset. Such a stance would have to have been directed from the top. Thus if, as Kachura and others stated, regional party leaders were hostile to the strikers, this could only have been an indication that the party hierarchy in Kiev was—as was to be expected—similarly opposed to the strike.

After the strike, the view of the striking miners were canvassed by sociologists from the Donetsk Scientific Center, which is affiliated with the Ukrainian Academy of Sciences.[61] Some 216 people from seventeen mines of the Donetsk Coal Association were interviewed. Of this number, 199 were workers and seventeen were engineers or technical workers. First, the respondents were asked to name the main reasons for the strike. Although the results were presented somewhat haphazardly, they appear to have been approximately as follows (in percentages, as the causes of the strike):

SUGGESTED CAUSE
Shortages of basic supplies—86
Low wages—79
Brevity of vacations—62
Inadequacy of pensions—56
High prices of supplies, unsatisfactory housing conditions, and poor relations with the administration—39–41
Poor working conditions—33
Lack of social justice—32
Poor medical services—25

Soap shortages were said to be the most serious. The fact that there were such deficits "in peacetime" had evidently elicited great anxiety among miners for the future of their children. O. Novikova, an economist at the Ukrainian Academy of Sciences' Institute of Economics of Industry, provided an analysis of the

61. *Radyans'ka Ukraina*, August 15, 1989.

miners' grievances regarding wages, the length of vacations, and pensions. In theory, she stated, miners' wages were relatively high, averaging 346 rubles per month, and the vacation time before the strike was twenty-four days, but could be extended to thirty-six for various reasons. Miners could obtain their pensions at the age of fifty; hence here also, in theory, there should have been no cause for grievance. When measured in terms of income per family member, however, the miners' wages were shown to be inadequate. Only 13 percent of those polled earned more than 200 rubles per family member; 14 percent earned between 131 and 150 rubles; 23 percent between 101 and 130 rubles; 25 percent between 71 and 100 rubles; and 11 percent between 50 and 70 rubles. In other words, almost 60 percent received less than 130 rubles monthly per family member, and this, at a time of rising prices, hidden inflation and black-market speculation for needed goods, was woefully inadequate.

Housing space was severely restricted, so that extended families were compelled to live together; this had caused painful overcrowding. Among those surveyed, more than 50 percent had less than nine square meters of housing space per family member; 23 percent had less than four meters; and 17 percent did not have a home of their own.[62] Concerning vacations, miners felt that those working in especially hazardous conditions were not being properly compensated with additional days off. Pensions were not considered adequate in view of the number of health problems among miners. Analyzing the somewhat nebulous cause of the strike described as "lack of social justice," Novikova commented that this did not operate automatically and that it was the duty of the trade unions to look after the miners' social needs. Had they carried out their proper functions, she wrote, the strike would not have taken place.

The miners' responses indicated a complete lack of faith in their trade unions: 91 percent felt that the unions were incapable of leading the strike movement, and 60 percent saw them as subjugated to the interests of the mine authorities. Although a new law on the rights of trade unions was in preparation, Novikova argued that steps had to be taken at once to reorganize the unions' activities so that they could protect the interests of the

62. *Izvestiya*, August 11, 1989.

working class. The miners evidently wanted to go further, however; 87 percent of those interviewed said that a law on strikes was necessary, that is, that the Soviet worker should have the right to strike.

Finally, the miners were polled about what should be done to improve the economic situation in the industry. A large majority (73 percent) favored the dismantling of territorial production association; 75 percent called for a reduction of the administrative apparatus; 41 percent thought the situation might be improved by the introduction of new technology in the mines; 38 percent wanted coal to be exported abroad for foreign currency; and about one-third called for the improvement in the social circumstances of miners and ecological conditions. Only a small minority (16 percent) felt that a transfer to full self-accounting in the coal mines would improve the situation. Thus along with the general discontent, there appeared to be a lack of faith in "radical" economic reform—a point suggested by Vinnyk at the Ukrainian party meeting.

THE AFTERMATH: POLITICIZATION OF THE COALFIELDS

In the Donbass, the economic nature of the miners' initial demands were soon replaced by political militancy. The nature of the strike changed, though one could argue that part of this transformation was already developing once the strike committees had effectively taken control of major centers. The miners' contempt for the existing party and government authorities was almost total. The old system had failed them and it had to be replaced with a new one. Until the new system was ready, the strike committees would remain in place as the only guarantee that justice would prevail in the coal mines. The national coal strike ended on July 24, 1989, therefore, but the committees did not disband. Instead, representatives of the strike committees and the production associations representing the cities of Voroshilovhrad, Dnipropetrovsk, and Donetsk oblasts, and of Rostov Oblast in the RSFSR, called a meeting in the town of Horlivka (Donetsk Oblast) on August 17, 1989.

By the time this meeting was in session, it was being termed a "conference." It included fifty miners' leaders and focused principally on the fulfillment of "the vital living demands" of the miners. An original thirty-eight basic demands soon grew to 138. More importantly, the meeting created a new organization: the Regional Union of Strike Committees of the Donbass (RSSKD).[63] The RSSKD soon formed a Coordinating Council, comprising thirty-one representatives from both the strike committees of the coal associations and the strike committees that had been formed in the cities. It adopted a provisional statute on August 19, and resolved that its center and base of activities should be Horlivka. Within a week, the RSSKD had begun to take action in the mining towns of Shakhtarske and Makiivka and announced that it could not work with the current party authorities in these two cities. In Shakhtarske, the city party committee was bundled out of its offices—which were promptly renamed "July 18" after the first day of the summer strike—and First Party Secretary V.D. Rozhin, who had been in office only a few days, was forced to make a humiliating retreat to a children's cultural center on the outskirts of the city. In essence, Shakhtarske, Makiivka, Horlivka and a number of other towns were now under the control of the RSSKD.

As early as August, preparations were being made for a second strike. A month's notice was given, after which, on September 20, ten days of preparation were to be made for a strike to commence on October 1, 1989. In the meantime, the RSSKD sent representatives to the founding congress of the Rukh in Kiev in early September. The move suggests that the RSSKD was contemplating whether to join the Rukh. If so, the reaction of the miners toward the Rukh was mixed. The representatives of the Voroshilovhrad strike committees resolved to withdraw from the movement which, they stated, was "the enemy and a menace to our struggle." While these miners agreed with the Rukh in theory, they were less than happy with what they perceived as extremist and nationalist elements present, who were warmly applauded at the Congress.[64] Perhaps one problem was that of fundamental differences between eastern and western Ukraine. Russian-speaking miners, whose militancy had inflamed the coal towns of the east, came face-to-face with a Ukrainianophone movement led primarily by writers and intellectuals. In addition,

63. *Robitnycha hazeta*, September 8, 1989.
64. *Radyans'ka Ukraina*, September 14, 1989.

these were early days and it was yet to become apparent that the two groups had, in fact, much in common.

The political activities of the miners continued, however, and forced a response from the authorities. One attempt to address the issues through traditional means was a plenum of the central committee of the trade unions of coal industry workers in Moscow in mid-September 1989. For the first time, many of the fundamental problems being faced by the miners, such as the alarming number of deaths each year from mining accidents, were addressed frankly. The RSSKD meanwhile held a meeting with journalists from the Donbass area. Its members claimed that their organization and its goals had been badly misrepresented in the Soviet press. The agreement between the strike committees and the government had not been well publicized so that many miners did not know its details. There had been no indication in the media of the organizational role of the RSSKD in getting work back on a normal footing. One of the worst offenders in this regard had been the newspaper *Robitnycha hazeta* for its story of September 8, 1989, entitled "Does the Union of Strike Committees Want to Take Power?"[65]

In Moscow, in late September 1989, representatives of the RSSKD were received in the Kremlin by L.O. Voronin, First Deputy Chairman of the USSR Council of Ministers, and a variety of problems in the coal industry were discussed at the subsequent meeting. Voronin assured the miners that the government had no intention of reneging on its side of the agreement reached on July 24 and that preparations to give economic autonomy to the mines of the major coal-producing basins were well under way.[66] On September 20, Volodmyr Shcherbytsky, the Ukrainian party leader was suddenly removed from the CC CPSU Politburo at the end of a debate on the Soviet nationalities question. The event, while very significant, may not have been related to the aftermath of the coal strike, though there is no question that Shcherbytsky's harsh attitude toward the strikers did not help his case. Gorbachev also visited Ukraine at this time and used some of his time there to try to appease the angry miners and avert a pos-

---

65. *Sotsialisticheskii Donbass*, September 23, 1989; and *Radyans'ka Donechchyna*, September 22, 1989.
66. *TASS*, September 27, 1989; *Robitnycha hazeta*, September 29, 1989.

sible strike on October 1. The miners, in their turn, demanded that the USSR Supreme Soviet address the issue of pensions without delay. Gorbachev told them that a strike was an extreme measure that would bring ruin and anarchy to the country's economy.[67]

On September 30, on the eve of the proposed strike, Mikhail Shchadov went on Soviet television to affirm that the main demands advanced by the miners had now been met. Before he did so, he made a lengthy speech about the dire economic effects of the original strike. He noted that because of the July strike, the country had lost over ten million tons of coal, including two million tons of valuable coking coal. The industry, the state, and the workers had all suffered because of the strike, he declared. The workers had a shortfall of some thirty million rubles in the social incentive funds alone. Implicit in his comments were that the striking miners were to blame for this state of affairs. Although resolution of all the demands from the various coalfields could hardly be resolved overnight since there were over 1,000 of them, Shchadov stated, some major steps had been taken in the direction of improving and building more housing, increasing wages and vacations, and cutting down on the bureaucratic and administrative apparati to give more authority to the enterprises themselves. A draft law on pensions had been submitted to the government.[68]

The RSSKD agreed to postpone a renewed strike, but this did not mean that it was satisfied with the existing situation. In mid-October 1989, some of its members went on television to complain bitterly that the various demands of the miners were being continually put off, particularly those concerning vacations and pensions. It insisted that a government commission be sent to Donetsk on October 17 to confirm that the miners' demands would be met without further delay.[69] Tension mounted in the mining regions when a Donetsk strike leader and member of the RSSKD, Aleksandr Sotnikov, was brutally murdered following an argument at a party, in the town of Zverevo in Rostov Oblast. Sotnikov had been conducting an investigation into corruption in

67. *TASS*, September 29, 1989.
68. Soviet Television, September 30, 1989.
69. Soviet Television, "Vremya," October 12, 1989.

the mines so that his murder was thought to be politically moti-
vated. His funeral was the occasion for massed solidarity of the
ranks of miners, who came from as far away as Karaganda for
the event.[70]
A year after the strike, the situation was still far from clear. A
renewed plea was made for increased investment in the Donetsk
Basin coalfield, this time by M.H. Chumachenko, Director of the
Institute of Industrial Economics of the Ukrainian Academy of
Sciences, who also had some new ideas about regional autonomy
in the coalfields.[71] In June 1990, miners across the country held
their first-ever congress, which took place over the course of five
days in Donetsk. It discussed the formation of an independent
trade union and the implementation of the No. 608 Resolution
of the USSR Council of Ministers on this same topic and on the
introduction of a market economy in the coal mining regions.
The mining delegates, however, had no interest in such discussions. Their view—and it was widely expressed—was that the
country simply needed a new government: "we demand the resig-
nation of the government. We urge the President [Mikhail Gor-
bachev] and the USSR Supreme Soviet to form a new government." A subsequent comment also provided a revealing postscript on the miners' attitude to the Communist Party that, they
believed, had so neglected them:

> The congress emphasizes the complete independence of work-
> ers' organizations from any political formations.... Our striv-
> ing for independence also determines our attitude toward the
> CPSU.... We demand that the CPSU be immediately deprived
> of its privileged status at our enterprises and institutions. The
> favorable service record for full-time Party and Komsomol
> functionaries must be cancelled. We believe that in conditions
> of a multiparty system the issue on the nationalization of CPSU
> property, created by the people, must be settled, and equal op-
> portunities must be created for all parties.[72]

One should bear in mind that there were other reasons why
the party had become so discredited, mainly related to the eco-

70. *Sotsialisticheskaya industriya,* October 21, 1989.
71. *Radyans'ka Ukraina,* January 27, 1990.
72. *Moscow News,* No. 25, July 1–8, 1990, p. 4.

nomic crisis in the Soviet Union. The miners, however, had become the most militant sector of the working community, and particularly in the Donbass. The wheel had turned full circle since the days of Stakhanov. In some ways, the achievement of the miners has been even more impressive than the growth of the Rukh in the midst of a repressive society dominated by the party from 1972 to September 1989. For members of the Rukh, after all, were nationally conscious intellectuals who had long rejected the party in practice even while retaining their membership.[73] The coal miners were, to all intents and purposes, depoliticized, apathetic, and russified to such a degree that they lacked even a knowledge of the Ukrainian language, the native language for a majority of them. Yet their organization and actions had been truly remarkable, perhaps in emulation of trade union movements outside the country, but certainly without relation to the official Soviet trade union movements.

As an indication of how far the miners had progressed, one can refer to a story cited by Kevin Klose in his book *Russia and the Russians*. In 1970, the coal miner Aleksei Nikitin had attempted to alert the authorities to miners' problems through the trade union of the Butovka mine in the Donbass. He was not successful and turned finally, in frustration, to Vladimir Degtyarev, the Donetsk Oblast First Party Secretary and reportedly received the following response: "If you stick your nose into our business, I'll mix coal with your blood and take your body and grind it into fertilizer."[74]

In conclusion, the coal strike of 1989 had distant and tangible origins. In retrospect there was an inevitability about it, but it required considerable courage on the part of the miners. Nevertheless, the results of that strike are uncertain. At the time of writing, the miners of the Donbass and Kuzbass regions had just carried out a repeat strike on the first anniversary of the July events that demanded an end to all party authority in the mining regions. The condition of the miners has improved only slightly. In the long term, it seems likely that some of the more dangerous mines must be closed down. The coalfield's future has gained a

---

73. Interview with Dmytro Pavlychko, Edmonton, Canada, June 26, 1990.
74. Kevin Klose, *Russia and the Russians: Inside the Closed Society* (New York: W.W. Norton), 1984, p. 65.

breathing space because of the popular assault on nuclear power in Ukraine, but one feels that the industry here cannot survive for long. The memory of the strike, however, will not be forgotten and it has formed an important landmark in Soviet history as the time when the workers first rose en masse against a self-styled workers' state. In Ukraine, it also revealed the widespread discontent in Eastern Ukraine, the heartland of the Soviet regime.

7 | CONCLUSION

WE HAVE ANALYZED A number of trends in recent economic and ecological development in Ukraine, with emphasis upon the public input into the changes in national life. Arguably, the era of the Communist Party is approaching its close, although its roots are perhaps more deeply embedded in Ukrainian society than is often recognized. Ukraine, once a great industrial nation though only briefly an independent one, has been ruthlessly denuded of many of its valuable resources, including its agricultural land. In addition, a great accident at Chernobyl continues to have a major impact not only on the economy of Ukraine, but also on the population itself. Early in 1990, a reporter for *The New York Times,* who was making a rare visit to the Ukrainian city of Chernihiv prior to the elections to the local Soviets and Supreme Soviet, remarked on what she called "widespread apathy" in Ukraine. People, she felt, did not care what happened during the elections. Such comments have been repeated so often that they cannot be ignored. Is it really the case, that in an era of dynamic change, Ukrainians are too apathetic to take a more active role in it?

I do not believe this to be the case. What the reporter misperceived as apathy was in fact an impression that recent changes have only been cosmetic. There is, it is true, an opposition within the Ukrainian Parliament for the first time and one that while a minority, is continuing to grow as party members tear up their cards. But the average citizen may no longer believe that any sig-

218

nificant economic or political change can be achieved by the present government. What is the point, they may ask, in taking part in a transition toward a more democratic process if the old system is to be retained; if the Union is to be reformed or the Communist Party is to reappear under a different name? They have lived, maybe, through the "seventy years on the road to nowhere," and in that time, while the totalitarian system may have loosened—some might say, even fallen—there is less food on the shelves than ever before, inflation is rampant and price rises render any wage increases meaningless. Some sectors of society therefore will vote for the Rukh, for the more democratic forces but without any real hope that the situation will change for the better.

In a conversation of 1990, Venyamin Sikora, one of the leading economists within the Rukh, maintained that the Ukrainian economy needed urgently to be transferred to a market system, but that this could only be achieved by means of substantial investment from the West. Much has been said on this topic. In 1990, the author attended two conferences largely on this subject: the first at NATO headquarters in Brussels, and the second at the University of California at Santa Barbara. There was agreement that joint ventures in the Soviet Union existed more on paper than in reality; there was also consensus that some of the East European economies—Bulgaria and Romania were cited—constituted too great a risk for foreign investors in the summer of 1990, and that a "wait and see" policy was advised. The same applies with even greater force to a future Ukrainian state, with the dual threat of civil strife and widespread hunger facing the population. Where is this investment to come from, and in what are Western businessmen likely to invest? And is foreign investment always the correct answer to the economic ills that befall an Eastern, quasi-socialist country?

Western statesmen have spoken darkly of "hard times" that are ahead for republics like Ukraine and pondered whether Ukraine and Belorussia might not form the heartland of the old Soviet state, along with Russia, and together form a new state as a future economic rival to the dominant Germany or the European Economic Community (though this hypothesis occurred before the above republics issued declarations of state sovereignty).

To some Ukrainians, in both Ukraine and the West, any thought of an economic or political union with Russia is regarded as anathema, and with some justice insofar as the russification and repression of Ukraine has traditionally been carried out by governments based in Moscow or what is today Leningrad. But in what does the future lie? Must one adhere to the gloomy and pessimistic scenario? We have examined the ecological and economic predicaments in Ukraine so can therefore ask: is there any hope for a brighter future in the sovereign Ukrainian state that was formally announced on July 16, 1990? Furthermore, what will be the likely political structure of such a state?

In the first place, we are likely to see a somewhat deindustrialized Ukraine, or at the least, a concentration on light and consumer industries rather than heavy industry. Even the Communist Party of Ukraine announced a moratorium on the construction of nuclear power plants in Ukraine at its Congress of Summer 1990,[1] thereby adopting a platform that had long been advocated by the opposition. Because of ecological damage brought about by past, centralized policies, future industrialization in Ukraine is likely to be very limited. One can project the closure of several major chemical and steel works, and the eventual abandonment of some highly dangerous coal seams. In taking such a step, the new government would only be recognizing the obvious: that because of the depletion of raw materials over the past decades, Ukraine will once again have to become, first and foremost, an agricultural nation. To act otherwise would be to ensure that Ukraine as a country becomes uninhabitable for decades.

Second, the new government will have to borrow heavily in order to complete the inquiry into the Chernobyl disaster, the elimination of existing problems from that accident, and the provision of adequate health care for the affected population. A common policy might be elaborated with the government of Belorussia, and possibly with Russia, the third most affected republic. This difficulty has at least been realized and the urgency of the problem has been made apparent. When Ukrainians or Belorussians declare that "Chernobyl is our tragedy," then they

---

1. *Robitnycha hazeta,* June 29, 1990.

use the latter word in the most literal sense as a catastrophe that has already affected the next generation and has demonstrated a lamentable and enduring legacy. No present or future government of Ukraine can dismiss the effects of Chernobyl or underrate their impact on the population, materially, physically and psychologically.

Third, the independent Ukrainian state will have to ensure the continuing progress of independent trade unions to protect the rights of workers. There are significant rights that have been attained under the old system—such as social insurance and health care—that are not guaranteed with a transition to a market or capitalist system. Such rights will be particularly important as the level of unemployment rises. At the same time, one can anticipate a decline in the influence of the Communist Party over the worker, though one cannot prognosticate the speed of that decline. It could occur virtually overnight, or else it could take years given the entrenchment of the old party apparatus in Ukraine. In the summer of 1990, the Strike Committees of the Donbass spoke out for the removal of all party cells in the coal mining industry. Other workers are likely to emulate their example. The Communist Party of Ukraine has lost the respect of the population, a respect that was largely based on fear: the stick rather than the carrot; and on the eminence of the CPU within the all-Union party structure. Even party members in Ukraine by the spring of 1990 overwhelmingly preferred an independent Ukrainian party organization over one that simply took its orders from Moscow.

Fourth, and this is a topic for the partner volume in this series, much will depend on the raising of Ukrainian national consciousness in Eastern Ukraine. One suspects that this process has already taken major strides through actions taken in Western Ukraine, and through the leadership of the Donbass miners in the strike of the summer of 1990. The development of the Ukrainian language and culture, which has proceeded at a great pace, is and will continue to be of paramount importance in Ukrainianizing the industrial cities of Eastern Ukraine (with the exception of Kiev). One can dispense with the Crimea, which is unlikely to remain part of a future Ukrainian state and was a Stalinist creation from the first. Many Ukrainians consider that the peninsula should be returned to the unfortunate Crimean Tatars, who were deported by Stalin during World War II. Odessa, which is likely to remain part of Ukraine, need not necessarily be subjected to the same Ukrainianization. Historically it has been a diverse and cosmopolitan city; plans are in motion to convert the entire city region into a "free economic zone."

Finally, it would appear that the agricultural Ukrainian state that will arise from the ecological ruins of Soviet Ukraine will require a close relationship with an industrial state. This is less likely to be Poland—an ecological disaster zone that bears a close resemblance to Ukraine in this respect—than Russia. For despite past history and animosities, Russia represents the best hope for the future economic survival of Ukraine, especially in terms of unexploited and underexploited natural resources. One need not suggest any form of subordination of Ukraine to a revitalized Russia. Quite the opposite. The supposition is that independent Ukraine might form an economic union with Russia that is based on equal prices for agricultural and industrial goods. Whereas iron ore and coal in Ukraine have only a few years of existence (perhaps thirty to fifty) and oil and natural gas are in short supply, Russia's reserves are plentiful and for the most part easily accessible. There is no question that there would arise a danger that the larger nation would again try to dominate the smaller. But one is surely hoping that democratic forces will have prevailed in both cases; that the days of "imperialism" have ended, though economic competition from other parts of Europe would be intense.

The above, then, is but a prognosis, and one that is purely speculative. The chances of economic failure remain very high for Ukraine. Moreover, all the measures suggested, from deindustrialization to the Chernobyl cleanup, require sums of money that the present government simply does not possess. Above all, the question arises whether Ukrainians have the unity and moral will to survive as an independent nation. Aside from the Ukrainian Republican Party, it is plain that all the non-Communist parties anticipated a very slow road toward independence: the decision of the parliament represented a surprise. Yet what would happen if the "center" collapsed first? Or if Russia broke away from the Soviet Union? Then Ukrainians would be forced to take some steps on their own. Members of the Rukh, in particular, have been among the most statesmenlike in trying to determine future policies based on such scenarios. But it would be naive to think that were such events to occur tomorrow, that they would have much hope of succeeding. Yurii Shcherbak's great fear—anarchy—would be the more likely outcome.

However, let us tread on firmer ground. Our analyses have shown that there is a definite and vibrant public spirit in Ukraine, and that this spirit has grown and flourished during the Gorbachev years, 1985–90. Public confidence has also grown. From total disillusionment at the ability of the popular will to bring about any change in economic planning, we now have a situation in which one cannot build a factory without republican approval, and one cannot acquire such approval without first asking the public for its support. This is a vast improvement upon the past. Further, while it seems trite to put the situation in these terms, it is clear that Ukrainians have shown great intelligence and ingenuity in analyzing their problems (though not the solutions) since the time that they were permitted to air their views publicly. Indeed, it could be argued that at times there has been rather more rhetoric than coherent action, but this is inevitable at a time when a nation is finding a voice that had been stifled for years. This voice is not always based on reason. Why should it be, given the repression of the past? It is sometimes extremist, sometimes too cautious, sometimes too emotional. But it is being heard, nonetheless.

Any future state that is economically and socially healthy will have to overcome, however, the economic devastation that is in place at the start of the 1990s. It is akin to rebuilding a nation after a long period of warfare, or even to constructing a society anew. It is more difficult for Ukraine than for the nations of Eastern Europe, which elicit more attention and sympathy from the capitalist West, where Ukraine the nation is still virtually unknown. To date, it is very hard to perceive any rational plans for economic development from the Ukrainian authorities or anything other than a vague hope that the West will assist in some way. Yet events such as Chernobyl or the case of the Chernivtsi children—paradoxically—have been instrumental in uniting Ukrainians, in forcing them to recognize that it is their fate and that of their country that is being threatened. Metaphorically the situation resembles that of the flooded house in D.H. Lawrence's book, *The Virgin and the Gypsy*. The flood has risen to the second storey of the house and the occupant has awakened, finally. There is time to get out of the house, but is there any hope of preventing the house, and home, from being washed away?

BIBLIOGRAPHY

SELECTED SOURCES: RUSSIAN AND UKRAINIAN

Atomnaya nauka i tekhnika (Moscow, 1987)
Ekho Chernobylya (Minsk)
Ekonomicheskaya gazeta (Moscow)
Ekonomika i zhizn' (Moscow)
Ekonomika Radyans'koi Ukrainy (Kiev)
Istoriya mist i sil Ukrains'koi RSR: Zhytomyrs'ka Oblast (Kiev, 1973)
Izvestiya (Moscow)
Komsomols'koye znamya (Kiev)
Komunist Ukrainy (Kiev)
Lenins'ka molod (L'viv)
Lesnaya promyshlennost' (Moscow)
Literaturna Ukraina (Kiev)
Literaturnaya gazeta (Moscow)
Literaturnaya Rossiya (Moscow)
Moscow News
Molod' Ukrainy (Kiev)
Narodne hospodarstvo Ukrains'koi RSR u 1987 rotsi: Statystychnyi shchorichnyk (Kiev, 1988)
Nash sovremennik (Moscow)
Nauka i suspils'tvo (Kiev)
Nauka i tekhnika (Moscow)
News From Ukraine (Kiev)
Novosti (Moscow)

Novyi mir (Moscow)
Pid praporom leninizmu (Kiev)
Prapor komunizma (Kiev)
Pravda (Moscow)
Pravda Ukrainy (Kiev)
Pravitel'stvennye vestnik (Moscow)
Rabochaya trybuna (Moscow)
Radio Kiev
Radio Moscow
Radyans'ka Donechchyna (Donetsk)
Radyans'ka Ukraina (Kiev)
Robitnycha hazeta (Kiev)
A.E. Romanenko, et al., *Meditsinskie aspekty avarii na Cherno-byls'koi AES* (Kiev: Zdorovya, 1988)
Sem'ya (Moscow)
Sil's'ki visti (Kiev)
Sobesednik (Moscow)
Sotsialisticheskaya industriya (Moscow)
Sotsiologicheskie issledovaniya (Moscow)
Sotsialisticheskii Donbass (Donetsk)
Sovetskaya Belorossiya (Minsk)
Sovetskii shakhter (Moscow)
Sovetskii sport (Moscow)
Soviet News and Views (Ottawa)
Stroitel'naya gazeta (Moscow)
TASS
Trud (Moscow)
Tvarynnystvo Ukrainy (Kiev)
Ugol' Ukrainy (Kiev-Donetsk)
Ukraina (Kiev)
Ukraine (Kiev)
Vechernyi Donetsk (Donetsk)
Vechirnii Kyiv (Kiev)
Visnyk Akademii nauk Ukrains'koi RSR (Kiev)
Visti z Ukrainy (Kiev)
Yunost' (Moscow)
Zelenyi svit (Kiev)
Zhovtnevi zori (Narodychi, Zhytomyr Oblast)
Znannya ta pratsya (Kiev)

SELECTED SOURCES: ENGLISH

Echoes of Glasnost in Soviet Ukraine, ed. Romana M. Bahry (North York, Ontario: Captus University Publications, 1989).

Henry Hamman and Stuart Parrott, *Mayday at Chernobyl* (London: New English Library, 1987).

Viktor Haynes and Marko Bojcun, *The Chernobyl Disaster* (London: The Hogarth Press, 1988).

David R. Marples, *The Social Impact of the Chernobyl Disaster* (London: The Macmillan Press; New York: St. Martin's Press; Edmonton: The University of Alberta Press, 1988).

Zhores Medvedev, *The Legacy of Chernobyl* (New York, W.W. Norton, 1990).

Bohdan Nahaylo and Victor Swoboda, *Soviet Disunion: A History of the Nationalities Problem in the USSR* (London: Hamish Hamilton, 1990).

Radio Liberty Research Bulletin.

Report on the USSR.

Yurii Shcherbak, *Chernobyl: A Documentary Story* (London: The Macmillan Press in association with the Canadian Institute of Ukrainian Studies, 1989).

| INDEX